# THE PHYSICS BEHIND...

# THE PHYSICS BEHIND...

**Discover the physics of everyday life**

FIREFLY BOOKS

RUSS SWAN

# A Firefly Book

Published by Firefly Books Ltd. 2018

Design and Layout Copyright © 2018 Octopus
Publishing Group
Text Copyright © 2018 Russ Swan

First printing

**Publisher Cataloging-in-Publication Data (U.S.)**

Library of Congress Control Number: 2018934785

**Library and Archives Canada Cataloguing in
Publication**

Swan, Russ, 1960-, author
      The physics behind... : discover the physics of
everyday life / Russ Swan.
Includes index.
ISBN 978-0-228-10089-8 (softcover)
      1. Physics--Popular works.  I. Title.
QC24.5.S93 2018      530      C2018-901160-2

Published in the United States by
Firefly Books (U.S.) Inc.
P.O. Box 1338, Ellicott Station
Buffalo, New York 14205

Published in Canada by
Firefly Books Ltd.
50 Staples Avenue, Unit 1
Richmond Hill, Ontario L4B 0A7

Printed and bound in China

First published by Cassell,
a division of Octopus Publishing Group Ltd
Carmelite House, 50 Victoria Embankment
London EC4Y 0DZ

Russ Swan asserts his moral right to be
identified as the author of this work.

All illustrations and layout design by Tilly
@runningforcrayons

**Publishing Director:** Trevor Davies; **Senior
Editor:** Pollyanna Poulter; **Copy Editor:**
Sarah Green; **Art Director:** Yasia Williams;
**Picture Library Manager:** Jennifer Veall;
**Senior Production Manager:** Peter Hunt

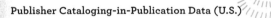

# CONTENTS

# INTRODUCTION

Life today is underpinned by scientific discoveries in ways that are not always appreciated. We take for granted that food can be heated in a few seconds, that aircraft will arrive at their destinations, that our news feed will be delivered to our pocketable device and that we can upload an image with the casual tap of a finger.

None of this would be possible without amazing breakthroughs in physics. In just four or five centuries, our understanding of the mechanisms of the Universe has moved from superstition and sorcery to our present, relatively enlightened, position. Humans have done this by being inquisitive, logical and sometimes a bit eccentric. Actually, quite often a bit eccentric.

It is in our nature to ask questions, to wonder how and why things work, to peek under the hood and mess about to see what happens. This is a quality that must be encouraged. Our collective tendency to mess with things has brought about great advances in science and engineering as well as the arts and humanities. It has also led us to do substantial damage to our home world, while giving us the insight to start putting things right.

This book gives a peek under the hood of the physics behind a range of everyday phenomena, from the devices we all rely on, to some of the more exotic stuff happening in the wider Universe. Rather than simply looking at what happens, we see how and why it happens.

We start with **modern life**, where commonplace technology is explored to see what makes it tick. Even mundane activities such as catching a train or weighing a bag of sugar involve some quite surprising phenomena, and have you ever wondered what really happens when you take a selfie?

Life would be dull without **entertainment**, and in this section we reveal the unexpected physics behind some of the ways we amuse ourselves. Why is a Blu-ray disc misnamed, and why was its diameter the result of something 200 years ago?

Everybody has some medical tests from time to time, but we rarely stop to consider how the **analysis** of our samples is carried out. The physics behind laboratory instruments, and some of the technologies arriving soon, may surprise you.

Rocket science used to be quite easy to understand, but recent developments in **space** are changing the very basis of exploration. Is there life elsewhere? What can physics tell us about the prospects for biology on Mars?

No questions are bigger than the questions posed by **big science**. We live in an age of unprecedented scientific discovery, where the biggest and smallest things in the Universe are being probed in ways never before attempted. Some of the answers seem so unlikely that they have a section all of their own: **Weird Universe**.

Back on Earth, we explore some of the physics behind **the environment** of our planet, including unexpected phenomena in the natural word and the realities of what human activity is doing.

**Transportation** is changing faster than many other technologies. Here, we explore the physics behind driverless cars, speed records and stealth. **Everything else** looks at a few oddities, including why every magnetic compass in the world is wrong, and what is really going on when you read this book.

In all this, we've tried to reflect the most up-to-date knowledge available. Knowledge changes, and scientific fact is sometimes just this year's best guess. Let's peek under the hood together, and explore the physics behind...

Russ Swan

# MODERN LIFE

# WI-FI®

If you ever feel like your Wi-Fi® connection has disappeared into a black hole, you might take some comfort from knowing that the technology only came into existence because of a failed physics experiment to build a black hole detector.

**W**i-Fi® is everywhere these days, with billions of devices including laptops, televisions, mobile phones and security cameras all using it to connect to each other and the wider internet. In essence it sends and receives digital data by radio communication between a device and a base station such as your home internet router (if you want to be really accurate, it uses parts of the electromagnetic spectrum in the microwave region, which are shorter and higher frequency than radio waves).

Getting digital information across relatively short distances with low power devices might seem to have little in common with radio astronomy, which uses huge receiver dishes to peer into the far reaches of the Universe. It might seem to have even less to do with a Hollywood actress, or a way to steer torpedoes, but the world of physics is full of surprises.

Austrian-born actor Hedy Lamarr fled to the USA shortly before the beginning of the Second World War, and became one of Hollywood's biggest stars of the 1940s and 50s. She was also an inventor, and in 1942 was granted a patent (along with composer George Antheil) for a way to prevent signals to radio-guided torpedoes being jammed. Her idea of "frequency-hopping" was not properly recognized at the time, but later became an important feature of many wireless technologies including both Wi-Fi® and Bluetooth®.

Thirty years later, physicist Stephen Hawking came up with the (then) astonishing theory that black holes might not be completely black, but actually emit some radiation. This should, in theory, allow them to be detected. Among those trying to find this predicted Hawking Radiation was electrical engineer John O'Sullivan, who reasoned that all that was needed was to make radio telescope images sharper and more sensitive. His idea involved using Fourier optics – see the "Deconstructing a chord" panel on page 12 – and would have been brilliant if it had worked. It didn't, and black holes remain invisible to this day, but the work did lead to patents for wireless networking – the now ubiquitous Wi-Fi®.

## Patent controversy

Major patents for Wi-Fi® technology were granted to CSIRO (pronounced sigh-roh), but with the huge commercial success of the technology there were many rival claims. In several US court cases, the Australians' patents were upheld and Dr O'Sullivan's claim to be the inventor of Wi-Fi® reinforced. The main patents expired in 2013, after earning hundreds of millions of dollars, but all those court cases mean that lawyers probably made more.

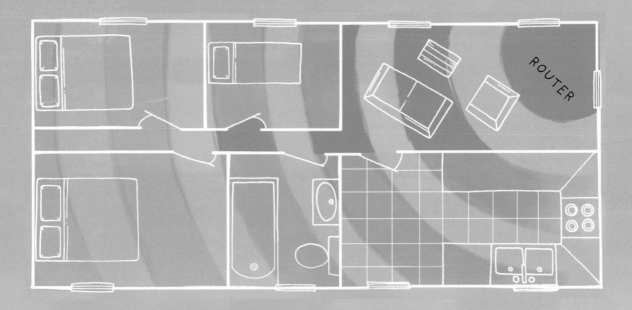

## The multipath problem

A clean Wi-Fi® signal has to overcome the problem of multipath interference, where the signal arrives at a device by many routes – directly from another device, or reflected from walls, ceilings and any other nearby reflective surface. Because of the way waves can combine or subtract from each other, the multipath problem could make practical Wi-Fi® almost impossible were it not for a couple of tricks.

First is Lamarr's **frequency-hopping** technique, which splits the signal across different wavelengths that each reflect differently. Second is the inbuilt error checking in wireless devices, which uses a **checksum** calculation to reveal whether the information received is the same as that sent.

## What's a checksum?

Let's say the data being transmitted was THE PHYSICS BEHIND. If we give each of the letters and spaces a numerical value (A is 1, B is 2, and so on, with 27 for a space) we can transmit the word as 20 8 5 27 16 8 25 19 9 3 19 27 2 5 8 9 14 4

A basic check would be to include information on how many characters had been sent. This would show if any had been lost or gained, but not whether any had been corrupted – you wouldn't know whether the phrase was THE PHYSICS BEHIND or THE MOON IS CHEESE because both are 18 characters long.

A simple checksum might be to add up the numerical values of the letters (20+8+5 and so on) to get 228. The receiving station does the same calculation, and checks whether it gets the same answer. If it does, there's a good chance the data has been transmitted successfully. If not, there is definitely an error and the whole string can be sent again. Real-life checksums are more complicated but follow this basic principle.

## Deconstructing a chord

The failed black hole detector that we discussed on page 10 was an attempt to distinguish feeble Hawking Radiation from all the other radio noise in the cosmos, picked up by radio telescopes such as Jodrell Bank in the UK. The resulting jumble of wavelengths is a bit like a musical chord, where many different notes are played at the same time.

The trick here is called a **Fourier transform**, which is able to deconstruct a complex wave into several simple waves. It's a bit like hearing a C major chord on a piano and working out that it must be composed of the individual notes C, E and G.

French mathematician and physicist Joseph Fourier came up with the process in the 1820s, about the same time as he discovered the greenhouse effect of the Earth's atmosphere.

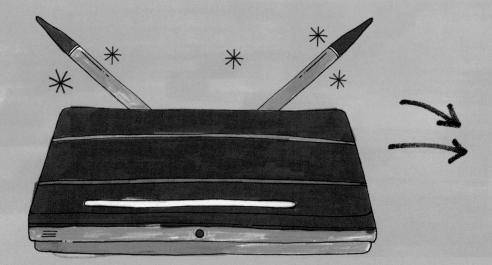

## Why is it called 802.11?

The term Wi-Fi® has become universal, and refers to devices based on standards known as IEEE 802.11.

IEEE is the Institute of Electrical and Electronics Engineers, a US-based professional association that has been operating since the 1960s. It has a number of committees, and wireless networking just happened to be the 11th standard agreed by committee number 802. Additional letters (802.11a, b and so on) are simply the various specifications of wireless networks.

The 802.11 standards are in a continual state of development and improvement, driven by the need to transmit more data more quickly and with greater security. Each new standard is marked by an alphabetical suffix, so 802.11g is slower than 802.11n. When about half of the alphabet had been used up, the designation changed to two letters: 802.11ac, 802.11ah and so on.

There's an old joke that a camel is a race horse designed by a committee, and the confusing list of wireless standards might be evidence that these were thrashed out by a series of committees. To be fair though, technology has improved rapidly in the late 20th and 21st centuries, and nobody could have foreseen how this would drive the demand for better faster Wi-Fi.

# Ever faster standards

The first 802.11 standard now looks rather quaint with its measly 1 or 2 megabits per second (Mbps) streaming capacity (think 5½ hours to stream a DVD – that's about three times as long as the movie itself).

To add to the confusion, the a and b standards were compiled at the same time, and it is sometimes thought that b came before a. This isn't quite true, but certainly 802.11a immediately demonstrated faster speeds as a direct result of using a higher radio frequency – 5Ghz instead of 2.4GHz. This created a new problem of its own, because the 2.4GHz band was already popular – and was being developed for future standards. The higher frequency offered more bandwidth but poor compatibility!

This problem is solved on later standards by using both 2.4GHz and 5GHz frequencies, allowing Wi-Fi networks from 802.11g onward to be compatible with most existing hardware. It's one thing to be able to stream content around the house at high speed, but the convenience is reduced if you have to replace the network card in your laptop and connected TV, and if your current smartphone can't talk to your router.

The latest standards bring even more speed and promise to maintain backward compatibility. Available from 2019, 802.11ax transmits at over 10Gbps, which is fast enough to download that DVD in just four seconds.

**SPEED** ⟶

|  | 802.11 | 802.11b | 802.11a | 802.11g | 802.11n | 802.11ac | 802.11ax |
|---|---|---|---|---|---|---|---|
| Frequency | 2.4GHz | 2.4GHz | 5GHz | 2.4GHz | 2.4 / 5GHz | 2.4 / 5GHz | 2.4 / 5GHz |
| Max speed | 2Mbps | 11Mbps | 54Mbps | 54Mbps | 100+Mbps | up to 1,300Mbps | up to 10.5Gbps |
| Performance | rather slow | compatible with g | incompatible with b and g | compatible with b | may interfere with b and g | compatible with b, g and n | available from 2019 |

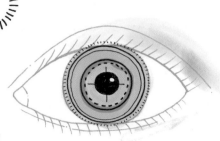

# FACIAL RECOGNITION

The smartphone is truly the wonder of our age, combining the power of a computer with an entertainment system, camera, and sensors to enable it to be anything from a spirit level to a video broadcasting station.

**W**ith all the value and personal data locked up in a small slab of glass and electronics, it's becoming more and more important to keep the bad guys out. This need will only increase as it becomes routine to use a phone as a wallet, paying for items in store with a simple wave.

One recent development that has rival phone makers trying to outdo each other is **biometric identification** – a big phrase which simply means checking that the person using the device actually is the owner. Biometrics are also used in passport control and other security applications, and seem certain to increase in future.

## Biometric ID

While a password or code relies on something you *know,* a biometric identifier relies on something you *are* – and so should be harder to crack. That's the theory anyway.

Lots of different metrics have been tried, with varying success. Fingerprint recognition is common, but may not be that difficult to hack. Iris scans, earlobe shape, voice ID and even pulse have been put forward and tried, and all can be made to work but may not be convenient enough for everyday use.

## Trust this face

In 2017, Apple introduced a new facial recognition system called TrueDepth, which uses an array of cameras, light sources, and sensors to create a 3D model of the owner's face and then compare this to anybody trying to unlock it.

The challenge of 3D is that it is really difficult to do with a single camera, but if facial recognition is only 2D it might be spoofed by something as simple as a photograph. TrueDepth cracks the problem by exploiting the properties of infrared light and the huge computing power now available in phones.

An infrared projector sprays a pattern of over 30,000 dots onto the face. When setting the phone up, a user must look at the screen and turn their head in a complete circle. The position of the dots is recorded by an infrared camera, positioned just about as far away from the projector as possible. Their apparent movement gives the depth perception needed.

This information is processed by an onboard **neural computer** that can, apparently, perform 600 billion calculations per second. That's about the same processing power as the best supercomputers in the world at the turn of the century. In your pocket.

## Fig. 1– Facial recognition

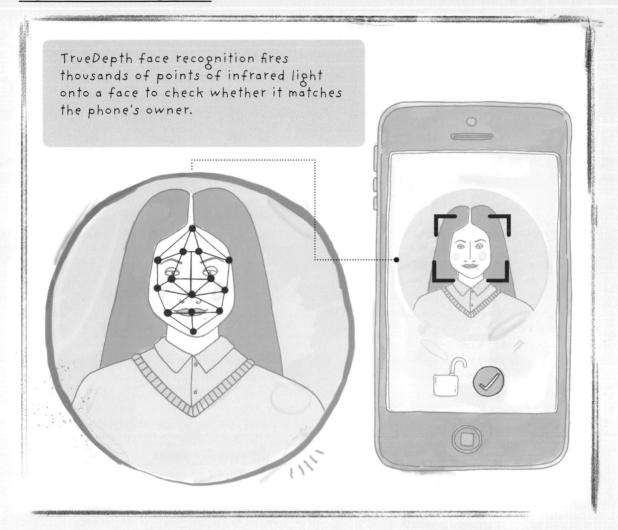

TrueDepth face recognition fires thousands of points of infrared light onto a face to check whether it matches the phone's owner.

## The learning machine

The importance of the neural computing approach is that this enables machine learning, in a way similar to a human brain. The neural processor used for TrueDepth should be able to tell if the person using the phone is you, regardless of whether you're wearing a hat, glasses, scarf, or are growing a beard. Neural processing is key to the most promising attempts at artificial intelligence.

Not everybody welcomes the growing presence of biometrics, but as the technology improves and becomes cheaper it seems certain that this will be a routine aspect of daily life in the future. One day we may expect to see DNA recognition as the "ultimate" in biosecurity, but it is certain that every increase in security is accompanied by a parallel improvement in the sophistication and cunning of would-be hackers.

# TOUCHSCREENS

The touchscreen has become the almost-ubiquitous interface between human and machine in just a few years. It can sometimes be a surprise to discover that a screen is not touch-sensitive, but merely an information display.

Touchscreens aren't all the same, and the technologies they use each exploit different physical phenomena.

## Resistive touchscreens

The variable electrical resistivity of a sandwich of materials, when subjected to pressure, was the trick behind early resistive touchscreens. Two conductive polymer layers are separated by a small air gap, and each layer is etched with a series of tiny parallel wires – on one layer going up and down, on the other left and right.

A small voltage is applied to the layers, but the electrical circuit is incomplete because of the air gap. Incomplete, that is, until a little pressure is applied to the top surface, causing it to deform and allowing the

two layers to touch. The etched conductive lines reveal the x–y coordinates of the point of contact, and tell the software what function you are asking for.

Resistive screens are considered a little crude nowadays, and can't really cope with multi-touch functions such as pinching and zooming, but they do have the advantage that they can be operated with a stylus or gloved hand.

## Capacitive touchscreens

More modern screens found on smartphones and tablets use a different electrical trick. These capacitive touchscreens are generally made of glass, which is more scratch-resistant than plastic, doped with conductive materials to hold a tiny electrostatic charge. This is the same type of charge you can get by rubbing a balloon on a woolly sweater, or walking over a synthetic carpet, only much smaller.

There are two slightly different ways that capacitive screens are made, called projection and surface capacitive, but in both cases a static voltage is applied across the top and bottom surfaces of the glass. When the top surface is touched by an electrical conductor such as your finger, even very lightly, the capacitance of the material changes at that point.

Capacitance is the ability of an object to store electrical charge. Making contact with the object causes the electrostatic field to change at that point. Sensors around the edge of the screen detect the position, and can cope with several points of contact simultaneously.

Touchscreens are the essential link in today's portable technology revolution, putting an incredible connected supercomputer in everybody's pocket. So we can look at pictures of cats.

### Fig. 1 – Resistive

When you touch the screen, the two layers meet, completing the electrical circuit.

## Fig. 2 – Capacitive

Your finger acts as an electrical conductor, changing the electrostatic field of the surface where you touch it.

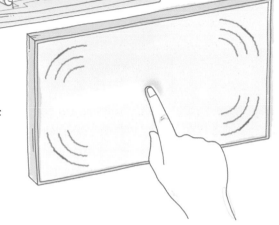

## Fig. 3 – FTIR

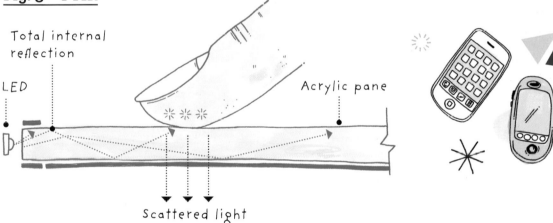

Total internal reflection

LED

Acrylic pane

Scattered light

## Next-gen touch

More precision and an even lighter touch is promised by a new generation of touchscreens, which exploit a property of light known as total internal reflection. This occurs when a beam of light is traveling toward a less dense medium (so from glass to air, for instance) at an angle greater than the critical angle for that material. On reaching the surface it is completely reflected back into the dense medium.

When an object approaches close to the surface, this total internal reflection is "frustrated" and some of the light is scattered instead of being reflected. The scattered light can be picked up by sensors on the far surface to reveal the location. **Frustrated total internal reflectance (FTIR)** screens can read fingerprints before the finger has touched the screen.

# MICROWAVE OVENS

Zapping a snack or a hot drink has become such a routine part of daily existence that it is easy to forget what a remarkable device the microwave oven is. If it was invented today, it would seem like something from science fiction: a small box that defrosts and cooks food using electromagnetic radiation, working in a fraction of the time of a traditional oven and using a fraction of the energy.

It's often thought that a microwave oven cooks from the inside out, while a conventional oven cooks from the outside in, but this isn't quite right. The microwaves heat up water molecules in the food as they reach them, and can only penetrate about 1.4cm (about half an inch) into the food. This means that anything less than 2.8cm (just over an inch) thick should be heated fairly evenly.

## Chips and dips

It's not quite that simple though, as the energy will be concentrated in certain areas and weakened in others due to the way microwaves reflect and combine. Think of a basin of water in which you are making waves with your hand. The waves will reflect off the sides of the basin and travel back toward you, meeting other waves as they do so. In some places the peaks will combine, making bigger peaks, and in others the troughs will combine to form deeper dips.

## Keeping it moving

The same effect happens with microwaves in an oven. They are generated by a **magnetron**, usually positioned behind the control panel, and bounce around inside the metal box until they meet a water molecule. They sometimes combine and sometimes cancel out, generating hot spots and cool spots. This is why the oven has a turntable – to keep the food moving for more even heating. It is also why some prepared meals give instructions to heat, stir and heat the food again.

Inside the oven, microwaves are reflected from metal surfaces and pass through materials like plastic and glass without much interaction. When they strike a water molecule, though, they cause it to vibrate and become hot. This heat is what actually cooks the food.

Because the energy in a microwave oven goes mainly into the water (and therefore the food), the cooking process is usually more energy-efficient than conventional cooking. On average, a microwave oven uses only a third to half the energy of an oven, saving electricity and $CO_2$ emissions as well as valuable time. Strangely, some foods such as porridge don't give this advantage!

# Making microwaves

The key part of a microwave oven is the magnetron, which applies **magne**tic force to elec**tron**s. A central negatively charged cathode is surrounded by a circular anode with several resonating cavities cut into it. As electrons are sprayed out from the cathode, they cross a magnetic field running at right angles and are accelerated into a spiral path.

Arriving at the anode, they are driven across the openings of the cavities. This causes oscillations in the electrical charge at the precise wavelength wanted – it's a bit like blowing air across the top of a bottle to produce a musical note at the resonant frequency. In this case the frequency is 2.45GHz, allocated by international agreements to microwave ovens.

## Fig 1. – Magnetron

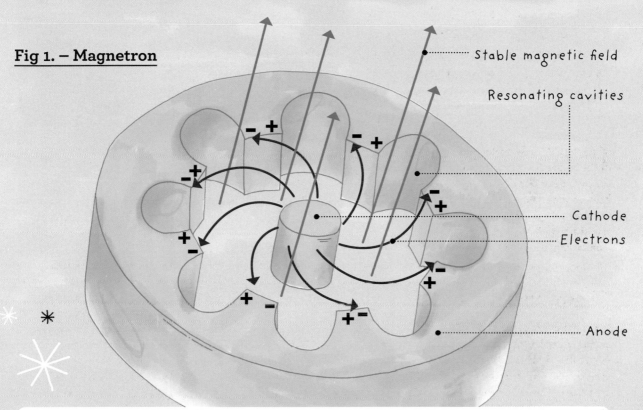

Stable magnetic field

Resonating cavities

Cathode

Electrons

Anode

## Accidental discovery

The microwave oven was invented in 1945 by Percy Spencer, an American engineer working on radar systems, which were a key Allied advantage in the Second World War. Spencer was working with a live radar set, and noticed that the peanut candy bar in his pocket was melting.

He experimented with other foods, making the first microwave popcorn and discovering that a whole egg would explode if placed in the radar beam – it splashed into the face of a colleague who was watching the experiment!

Spencer worked for Raytheon, which launched the first microwave oven in 1947. The RadaRange was the size of a large fridge and cost $5,000 ($68,000 today). It was a commercial failure, and poor old Spencer was paid a bonus of just $2 for an invention that changed the world.

Modern Life

# ICE CREAM AND BEER DIET

Great news for anybody trying to lose weight!
You can eat as much ice cream and drink as much
cold beer as you like because they are calorie-negative.

The basic idea behind many diets is to put fewer calories into your body than you burn up through staying alive – breathing, exercising and reading books. If the energy going in is less than the energy going out, you must lose weight. It's science.

## Cold comfort

The calorie is not officially recognized as a unit of energy, but despite this it remains widely used and understood. The most common definition is that one calorie is the energy needed to raise the temperature of one gram of water through one degree Celsius.

And this is where it gets interesting. Your home freezer will store ice cream at about –18°C (0.4°F), but it may well have warmed a little by the time you eat it. Let's say it's at –10°C (14°F).

## Body heat

Your body's internal temperature is 37°C (99°F). As you digest the ice cream, your body will warm it up to this temperature, using energy to do so. A modest 100g portion absorbs 3,000 calories this way (ice cream has a lower specific heat capacity, taking less energy to heat or cool, than water).

But a 100g serving of ice cream provides only about 300 calories. Woo-hoo! It seems you can eat ice cream all day. Let's wash it down with a cold beer from the fridge.

A 330ml (nearly three quarters of a US pint) bottle of good beer might deliver 130 calories. You drink it at about 4°C (39°F), and again your digestive system raises it to body temperature.

330g × 33 = 10,900 calories to warm it up internally, but only 130 calories have been consumed. This just gets better and better!

**Sadly, this is all nonsense.** There's nothing wrong with the principle, and the numbers are all about right. But the calorie, that "widely used and understood" unit of energy, is actually widely misunderstood. When somebody talks about the calories in food, those are kilocalories and will be marked on packaging as kcal in the UK (or just calories in the US). Using the definition above that one calorie is the energy needed to raise the temperature of one gram of water through one degree, that ice cream actually delivers about 300,000 calories to your body and needs 3,000 to be digested. You've still consumed 297,000 calories.

What this really shows is the value of having standardized units that are universal and unambiguous. Something like the International System of Units (SI), where the unit of energy is the joule not the calorie.

## Negative-calorie food

A popular modern myth is that some foods, particularly celery, require more energy to consume than they provide in nutrition. The negative-calories claim has even been used in advertising by major brands, and in pseudoscientific fad diets.

Sorry, but it's just not true. There are some foods and drinks that are almost zero-calorie, but even "calorie-free" soft drinks will usually provide some marginal nutrition (labeling laws permit this).

Water provides no calories, but absorbs some body heat and so provides a very marginal negative calorie effect. A litre of ice-cold water would roughly cancel out the calories in a single sandwich biscuit.

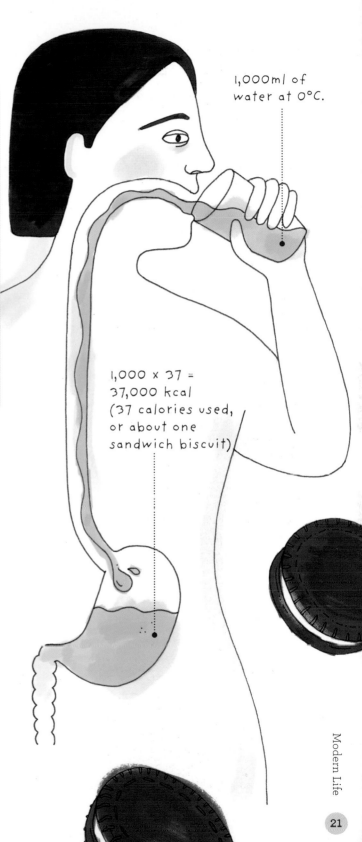

1,000ml of water at 0°C.

1,000 x 37 = 37,000 kcal (37 calories used, or about one sandwich biscuit)

37°C

# TAKING A SELFIE

It has become such a routine and mundane act that it is easy to overlook the startling science behind digital photography. You whip out your phone or camera, point it at yourself or your cat, and before you know it you're collecting likes and comments on the social media platform of your choice.

The simple process of taking a picture involves conventional physics that would have been understood by Isaac Newton, semiconductor science that might have puzzled Albert Einstein and the weird world of quantum mechanics. All built into something you carry around in your pocket.

The first part of the process is the most straightforward. Your camera or phone lens collects light from the scene, focusing it onto a sensor. When light enters a lens, it encounters a denser medium than the air and consequently slows down. This slowing causes the light rays to bend or refract, and the shape of the lens is designed to control this so that they all converge (focus) on the plane of the sensor.

One difficulty here is that different wavelengths of light refract by different amounts. This means that cheap single-element lenses may produce coloured fringes around images. For this reason, camera lenses are made of a number of separate optical components, each playing a part in maintaining a clean sharp image with true colours.

## Silicon sandwich

Once focused on the image sensor, the photons encounter a sandwich of two semiconductor materials. Both are primarily silicon, but are "doped" with carefully controlled impurities – often phosphorus and boron. Combining these elements with silicon results in a P–N diode. The P semiconductor has positively charged electron holes (as boron has fewer electrons), while the N has negatively charged excess electrons (due to extra electrons from the phosphorus). P–N diodes allow electrons to flow easily in one direction, from N to P, but not the other way. If a reverse current is applied, something quite extraordinary happens to the diode. It repels any electrons attempting to cross, and the sandwich exhibits the **photoelectric effect**.

This happens when a photon arrives in the silicon sandwich, knocking an electron out of its orbit and creating a new electron/hole pair. The charge in the silicon draws electrons up to the surface where they create a voltage proportional to the number of photons striking the sensor.

## CCD or CMOS

The two main types of image sensor both work on the same photoelectric effect, and differ mainly in the way they handle the information created. A CCD (charge coupled device) extracts the data row by row, in a relatively slow process that takes a lot of battery power. The images it produces are generally high quality.

The main rival is CMOS (complementary metal oxide semiconductor), which takes some of the data processing onto the sensor itself. This is a cheaper, faster and less power-hungry sensor which is said (by the people who make them, so they would say that) to now produce images of quality equal to CCD. Most modern digital cameras use CMOS sensors.

With the sensor divided into a grid of tiny squares, the picture elements (pixels) each produce a different voltage. This information is processed to create a binary digital file that you can upload to the web within moments of the picture being taken.

## Fig 1. – Photoelectric effect in a digital camera

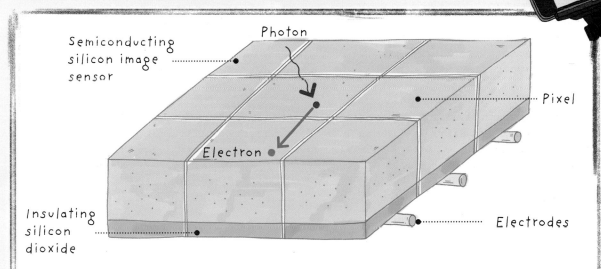

Semiconducting silicon image sensor

Photon

Pixel

Electron

Insulating silicon dioxide

Electrodes

Electrons form a tiny electrical signal, which is processed into a digital image row-by-row and pixel-by-pixel.

# FLASH MEMORY

Modern microelectronics seem almost magical in the way they work, with nothing to see and no moving parts. There is some very weird but incredibly useful physics going on at the tiny nanometre scale they are built on, involving force fields and materials that can't quite decide whether they are conductors or insulators.

The memory card in your digital camera or phone, the USB drive to transfer files and the solid-state hard drive (SSD) on a lightweight laptop all make use of flash memory. This is a particularly useful technology that can store a lot of data in a tiny space, find it again quickly and be re-written many thousands of times. It is also cheap to mass produce, robust and doesn't need a continual power supply to retain its memory.

Did I mention that it seems almost magical?

Flash memory, like the majority of microcircuitry, is based on a variation of the transistor. This is a simple device, usually having three electrical connections. When a current is applied across one pair of terminals, a current can also flow across another pair – so a simple transistor is a bit like a remote control switch.

Transistors rely on the properties of semiconductor materials, which behave in unusual ways. Some allow electricity to flow one way but not the other, or change their resistance according to temperature or light. Careful tweaking of the properties of semiconductors is achieved by the addition of tiny amounts of chemical impurity, or dope.

## The Moore the merrier

The integrated circuits (chips) in electronic technology contain many millions of transistors per square millimetre. These are made by printing microscopically-small dots, lines and sheets of conductor and semiconductor, then etching away unwanted areas, in a fabrication plant or "fab". The number that can be squeezed onto a chip has increased exponentially, doubling about every two years, following a growth curve known as Moore's Law.

Transistors exploit the strange properties of two differing semiconductors. Electrons flow between the terminals called the source and the drain, and in doing so change the behaviour of the semiconductor at the third terminal, the control gate. In a flash transistor, a second semiconductor called the floating gate stands as a barrier to the control gate, and is itself electrically insulated from the other components.

The insulation can be overcome by the force of the field effect, where an electrical field pushes electrons across the barrier. They become trapped in the floating gate, and cannot escape until another current draws them back across the insulator. When the floating gate is charged with electrons, it corresponds to a binary 0. When the electrons are withdrawn, the gate has a neutral charge and is a binary 1.

# How much does a digital photo weigh?

A brand new (or newly-formatted) flash drive has all its binary digits (bits) set as 1s, and data is written by changing some of them to 0s. This involves driving electrons into the floating gate, nominally increasing its mass by a tiny amount.

Each bit change from 1 to 0 involves about 1,000 electrons. Multiplying this up, an average 5MB digital photo has a mass of about 25 femtograms ($25 \times 10^{-18}$g), or about the same as a single virus, and a "full" 64GB memory might change its mass by roughly 250 nanograms, or about one millionth the weight of a single grain of pollen.

## Fig. 1 – How a Flash memory floating gate works

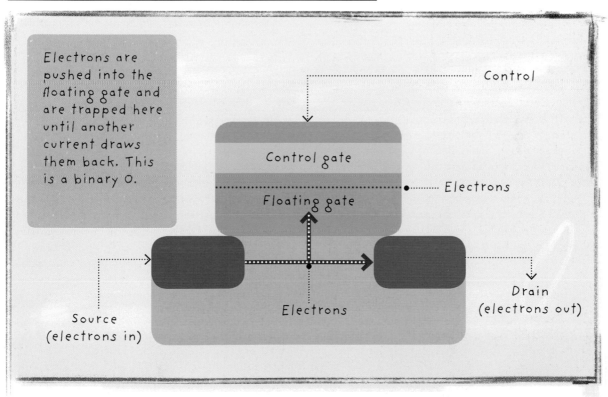

Electrons are pushed into the floating gate and are trapped here until another current draws them back. This is a binary 0.

Control

Control gate

Floating gate

Electrons

Electrons

Source (electrons in)

Drain (electrons out)

# A BAG OF SUGAR

For many people, the easiest way to visualize a kilogram is to think of a bag of sugar – but how do we know that the sugar inside the bag is actually the weight it claims on the label?

Standardization of units of measurement is vital for business and for scientific experiments (see **Ice cream and beer diet,** page 20 for example). Standards were originally based on physical objects, such as a ruler or a weight, but are now nearly all based on principles of physics. In fact, the kilogram is the last metric unit to be defined by an artefact rather than science.

## An electric kilogram?

The international prototype kilogram is a lump of platinum-iridium alloy kept inside three glass bell jars in a vault in Paris. Every other standard kilogram in the world must be calibrated against the "le grand K". Although this item seems to have mysteriously lost about 50 micrograms over a century, officially it remains precisely one kilogram.

But its days are numbered, and a new definition will eventually be adopted. Two methods were tried, one based on counting atoms and one that is known as the electric kilogram.

Counting atoms is simple to understand, but difficult to do. The idea is to use pure silicon, which has atoms arranged in a nice regular crystal lattice, and the universal constant of **Avogadro's number** – which is the number of atoms in one mole of a substance. A mole is the amount of a substance that has as many atoms as there are in 12 grams of carbon-12.

## Avogadro on toast

This method needs an accurate measure of the Avogadro number. It's reckoned to be about $6.022 \times 10^{23}$, but different experiments produce slightly different results and there's nothing a scientist likes better than the chance to claim their results are better than everybody else's.

The electric kilogram may be more promising, even though it also seems a bit strange. It is based on the **Planck constant**, and uses a precision weighing scale called a Kibble balance (or Watt balance).

The idea is to define the kilogram by balancing forces. Electric current in a coil is adjusted until a kilogram mass is balanced, and the result used to calculate the Planck constant.

This new kilogram, like the other SI (International System of Units) base units, could be defined in an email and reproduced by a competent scientist anywhere in the universe. That's more than can be said for a physical lump of metal.

Avogadro number:
$6.022 \times 10^{23}$

Planck constant:
6.62607004
$\times 10^{-34} m^2 kg/s$

# A metre measured by the speed of light

The metre was defined in 1791 as one ten-millionth of the distance from the North Pole to the equator, measured through Dunkirk. An expedition lasting seven years came up with the measurement – but made a small error, and the first standard metre ruler was short by about 0.2mm.

A more scientific definition came in 1960. It was declared that a metre would be 1,650,763.73 wavelengths of the specific orange-red light emitted by an atom of krypton-86, in a vacuum. Handy!

The slightly less cumbersome definition (since 1983) is based not on the *wavelength* but the speed of light. This, as far as anybody knows, is a universal constant and the upper speed limit of the Universe. A metre is officially the distance traveled by light in 1/299,792,458th of a second. That's great, but means you need an accurate second to measure an accurate metre (see **Magnificent seven** below).

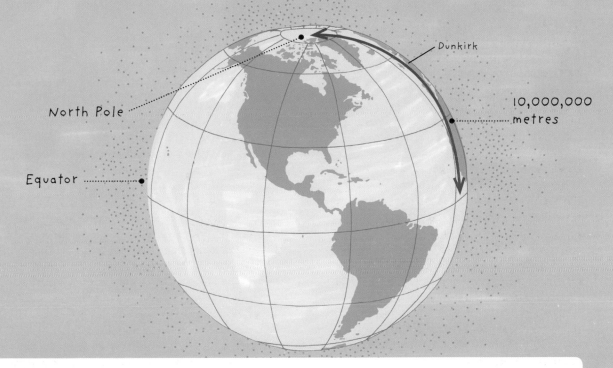

Dunkirk

North Pole

10,000,000 metres

Equator

## Magnificent seven

The metre, the second and the kilogram are the most familiar of the seven base units of the SI system. These and another four – electric current (ampere), temperature (kelvin), the unit of substance (mole) and of luminous intensity (candela), allow any known phenomenon to be quantified.

All SI units can be broken down to these seven fundamental measurements. The unit of power, for example, is the Watt. This is one joule per second, and a joule is the work done (the energy transferred) when a force of one newton acts over a distance of one metre. The newton is the force needed to accelerate one kilogram of mass at one metre per second per second.

Describing a motor as having a power of 100 Watts is much easier than saying it has the power of 100 kilogram metres squared per second cubed.

# CATCHING THE TRAIN

Before the newly-invented railways began connecting major towns and cities in the 19th century, each place might have its own standard time determined by a clock on the town hall. It wouldn't matter if two places several miles apart were on slightly different times. Most people did not have watches, and they organized their day around sunrise and sunset.

The introduction of the railway timetable changed all that, and meant that a wider standard time became necessary. Trains were able to distribute the official time as they went, and the whole concept of punctuality took on a modern meaning. Oh happy days.

The standard unit of time is the second, initially based on the spin of the Earth – one second was 1/86,400th of a day. The more scientific modern definition is a lot more precise and reliable, but hardly easy to remember. It is, officially:

> "The duration of 9,192,631,770 periods of the radiation corresponding to the transition between the two hyperfine levels of the ground state of the caesium-133 atom."

If you want to be on time for a train today, though, you won't have to worry about that because your wrist may be adorned with an inexpensive but impressively accurate quartz watch. This marvel of microelectronics and engineering has made us all take super-accurate timekeeping for granted, where once it was the preserve of the rich.

## Chronometry on the cheap

Watch enthusiasts reckon that a modern certified mechanical chronometer (fancy name for a fancy watch) will be accurate to about three seconds per day. However, a cheap modern quartz watch may be accurate to one second per day, for just one thousandth of the cost. So how does a quartz watch keep such good time, so inexpensively?

The timekeeping is down to the **piezoelectric effect**, a remarkable property of crystals and some other substances. When these are squeezed or otherwise manipulated, they generate a small voltage. What's even more useful is that the process works both ways, and the piezoelectric crystal responds to electrical stimulation by vibrating.

Quartz timepieces are regulated by the piezoelectric oscillations in a (you guessed it) quartz crystal. The crystal is tiny – just a few millimetres long – and shaped like a tuning fork because this allows it to vibrate steadily.

A small voltage from a battery stimulates the crystal making it vibrate exactly 32,768 times per second. Here's the really clever bit: it is vibrating because of the piezoelectric effect, and the piezoelectric effect means those vibrations generate electricity. Regular DC (direct current) electricity goes in, and a precise electrical beat comes out.

The rest is simply a matter of counting the oscillations and displaying the result in digital or analogue form. That is easy and cheap for mass produced microelectronics.

# Piezoelectricity

The piezoelectric effect is so useful that large parts of modern life have come to depend on it. One early application was in sonar systems to detect submarines, using a piezo transducer to produce ultrasound waves and detect any echo.

Probably the best known application is in the spark generator on gas stoves and barbecues, where a simple push makes a spark strong enough to cause ignition.

The property has also found use in record players, microphones and scientific instruments such as electron microscopes. There have even been attempts to use the piezoelectric effect to harvest energy from everyday activity – your shoes could become portable generators to recharge your phone while you walk, or run, if you are late for that train.

## Fig. 1 – Inside a quartz watch

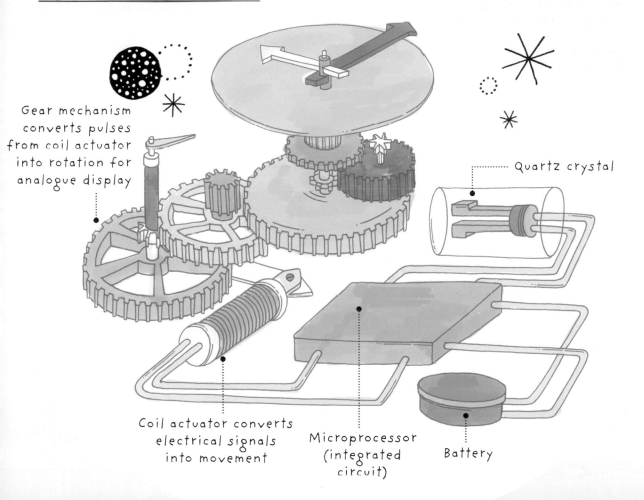

Gear mechanism converts pulses from coil actuator into rotation for analogue display

Quartz crystal

Coil actuator converts electrical signals into movement

Microprocessor (integrated circuit)

Battery

# CALENDARS AND CLOCKS

Much as we love planet Earth, its inbuilt arithmetic is rather poor. Its three basic natural measurements of time are the day, month and year, but none of these are a convenient multiple or fraction of each other. Scientists have devised ways to cope with this.

## Taking a leap

The standard calendar year is 365 days long, and 366 on leap years which are every fourth year. Everybody knows that, right?

There are two reasons for inserting a leap day. One is that the Earth actually takes about 365.242 days to orbit the Sun, and if we didn't make the adjustment the calendars would go out of synch with the weather. Nobody wants their August holidays in the winter!

The other reason is that leap years are also Olympic years, and this allows athletes an extra day in February to train. Okay, that's not true. But it could be.

Observant readers may notice that a solar year is "about 365.242 days". If you want to be really accurate, it's 365.242199074 days. This is not a nice round number, and so the simple extra quarter day would still cause calendars to drift. For this reason, a leap year is every fourth year except every hundredth year, except every fourth century (1600 and 2000) when it is. I hope that's clear.

Without any adjustment, the seasons would shift by 24 days each century. With a leap day every fourth year, the error would be almost a day per century. The additional century year rule means it will take over 3,000 years before the error amounts to a whole day.

## Take a little leap

While the leap year compensates for the awkward length of the solar orbit, the leap second is a very different beast. This is an additional second inserted into the clock to compensate for the gradual slowing down of the Earth's spin.

The official time of planet Earth is UTC, Universal Time Coordinated, which is based on the Sun crossing the Greenwich meridian. But the spinning Earth is a poor timekeeper, so we also have International Atomic Time (TAI, from the French *temps atomique international*) for when more precision is needed. This is based on super-accurate atomic clocks.

## Drifting in time

UTC and TAI have a tendency to drift apart and, when the discrepancy gets close to a second, a leap second is inserted to bring them back together. In effect, all the world's clocks stop for a second at midnight UTC on either 30 June or 31 December, allowing the planet to catch up.

The first leap seconds were added in 1972, and there have been a total of 27 since then. Before 2000, a leap second was added in 21 out of 28 years (75 percent), but since then only in 5 out of 18 years (28 percent). It seems that the planet has improved its punctuality.

# The physics of being middle aged

Whether you are middle-aged or not has a bit to do with biology, and a bit to do with psychology. But it might have a little to do with physics as well. It's all about how you count time.

A thousand seconds, a kilosecond, would pass in 16 minutes 40 seconds. A megasecond, a million seconds, is about 11½ days. How long is a gigasecond? It's a billion seconds, or about 31.7 years. Your first gigasecondiversary is about eight months and eight days after your 31st birthday, your second about four months and 16 days after your 63rd birthday, and your third about 24 days after your 95th.

If you know the time of day you were born, you can work out exactly when to celebrate – but remember to allow for leap years, time zones (if you've moved), daylight savings and leap seconds.

The gigasecond could define ages in humans: up to 1Gs, you are young; between one and two you are middle-aged; over two is senior and over three is pretty amazing.

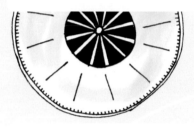

South African freedom fighter and politician Nelson Mandela and US astronaut John Glenn both lived beyond three gigaseconds.

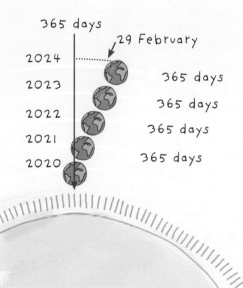

365 days

29 February

2024

2023  365 days

2022  365 days

2021  365 days

2020  365 days

Earth's yearly orbit of the Sun is not an even number of its rotations around its axis (days). After four years, an extra day allows the calendar to catch up to the orbit (but not every fourth year is a leap year).

# ENTERTAINMENT

# OPTICAL DISCS

Playing a movie on a home television involves little effort beside slipping a disc into a player and pushing a few buttons, but some seriously impressive physics makes this possible. Laser optics, quantum mechanics and nanotechnology all play a part in getting the latest instalment of *Star Wars* onto your screen.

It is quite remarkable just how much information can be stored on a small circle of polycarbonate plastic. The first discs in the standard 120mm format were CDs (compact discs) and were introduced in the early 1980s. These could store an impressive (for the time) 650MB of information, and became popular for distributing music and software.

Computers at this time used 5¼-inch floppy disks (slightly larger than a CD), which stored just 720kB. CDs offered a nearly thousand-fold increase by abandoning magnetic storage in favour of optical.

The information on a CD is encoded as binary data (ones and zeroes) stored as a series of tiny pits in the reflective metallic layer. You might expect that the pit would represent a one or zero, and the unpitted surface (called the land) the other, but actually it is the change between them that is detected. If the disc reader sees

"land – land – land" or "pit – pit – pit", it outputs zero – zero – zero. But "land – pit – pit" would be zero – one – zero (no change – change – no change).

The pits on a CD are about 800 nanometres across (roughly one hundredth of a human hair) and are arranged in a spiral from inside to out. If it were unravelled, the pit track would stretch over 5km (3 miles).

The once-impressive capacity of the CD was soon overtaken by demands for more content to be available in this pocketable format. To get the huge data file of a movie onto a disc meant using some clever physics.

Making the pits smaller is the obvious answer, but presents a problem. The smallest spot that light (laser or otherwise) can be focused to is a function of the wavelength and the aperture of the lens, and is known as the Airy disk.

|  | Laser wavelength | Spot size | Pit size | Track pitch | Data capacity |
|---|---|---|---|---|---|
| CD | 780nm | 1600nm | 800nm | 1,600nm | 0.65GB |
| DVD | 650nm | 1100nm | 400nm | 740nm | 4.70GB |
| Blu-ray™ | 405nm | 580nm | 150nm | 320nm | 25.00GB |

Airy disk diameter = 2.44 × wavelength × aperture (f-number)

The red laser used in CD readers has a wavelength of 780nm and can be focused to a spot 1,600nm wide. To cram enough data for a movie onto a DVD means using pits 400nm wide on tracks 1,100nm apart, making the red laser useless. So an orangey-red wavelength was selected, at 650nm.

The situation repeated with high-definition Blu-ray™ discs, which use an even shorter 405nm wavelength. This allows 25GB of information – nearly 40 CDs-worth – to be encoded on each side.

What colour is the laser in a Blu-ray™? It's not actually blue, but violet. A true blue laser would be around 475nm, but perhaps "Violet-ray" didn't sound good to the marketing people.

## Beethoven designed the Blu-ray™

It might seem unlikely that Ludwig van Beethoven, who died in 1827, was involved in the specification of modern optical discs – but this is just what happened. The original design for the CD was 115mm diameter, which would play 60 minutes of music.

According to legend, the president of Sony, Norio Ohga, demanded it be capable of playing an entire opera or all of Beethoven's 9th Symphony. The disc diameter was increased to 120mm, allowing up to 74 minutes of music to be stored.

DVDs and Blu-rays™ maintained the 120mm size, which would probably delight old Ludwig.

## Fig. 1 – Wavelengths, laser spots and pits

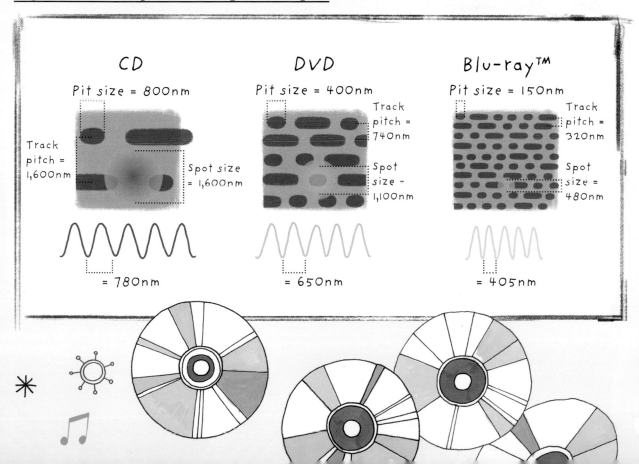

# LASERS

As we saw on page 34, the key component in an optical disc player is the laser. These specialist light sources were invented as recently as 1960, and even though they are now common and inexpensive, they are still pretty remarkable.

A perfect laser produces light of a single wavelength, not just red or blue but specifically at (say) 650nm. It does this by adding energy to a **gain medium**, which is trapped inside an **optical cavity**. A gain medium might be a crystal like garnet or ruby, a gas such as argon or carbon dioxide, or a semiconductor. It's often a mixture of materials.

The optical cavity is a small tube with a mirror at each end. One end will be a complete mirror, while the other will have a transparent section to allow the laser light to escape. When energy is added, by electricity or light, the atoms of the gain medium are stimulated. Electrons absorb energy, and move to a higher orbit around the atomic nucleus. Because of **quantum mechanics**, they can only move up or down by a complete orbit – there are no in-betweens.

## Out of orbit

The higher orbit cannot be maintained for very long (a few milliseconds) before the electron falls back. As it does so, it has to lose energy and so emits a photon (the fundamental particle of light). It is a characteristic of each gain material that the photons created are of a specific energy, giving a specific wavelength of light.

So far what we have is a bit like a fluorescent tube or neon sign. It makes light, but it isn't a laser. The real trick is the optical cavity, which confines the photons forcing them to reflect back and forth inside. As they do so, they sometimes collide with another atom.

## Quantum leap

This is enough to tickle that atom to perform the same quantum leap, releasing a new photon *in addition to the original one*. The light has been amplified by stimulating the emission of a new photon – which is where the name comes from: Light Amplification by the Stimulated Emission of Radiation.

These photons are not only the same wavelength, they are also synchronized so that their peaks and troughs line up. This **coherence** is what really distinguishes laser light from monochromatic (one colour) light produced by something like an LED. It also allows a laser beam to remain highly **collimated** (the rays of light stay parallel instead of spreading out).

A perfectly collimated laser would produce the same size spot at any distance, but in the real world beams will always diverge a little. A good laboratory-quality laser may have a beam divergence of only 0.5 milliradians (one milliradian is a thousandth of a radian or about 0.06 degree), which doesn't seem like much. Even so, a 1mm (0.04 inch) diameter beam would grow to about 5cm (2 inches) over a distance of 100m (328ft).

Divergence is not a big problem in optical disc players and cutting machinery, because the distances are quite small, but it does limit how far your laser pointer will produce a sharp spot.

## Fig. 1 – Photons pump back and forth within the laser cavity

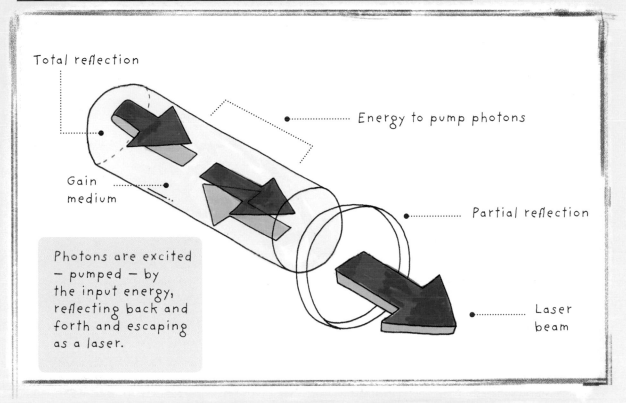

Total reflection

Energy to pump photons

Gain medium

Partial reflection

Photons are excited — pumped — by the input energy, reflecting back and forth and escaping as a laser.

Laser beam

## Measuring up a quantum leap

The term quantum leap has come to mean some huge and profound change in something, but in physics it is really the exact opposite.

When an electron moves from one orbit to another, it performs a quantum leap and either absorbs or releases a photon. One photon is the tiniest packet of energy that exists, and one electron orbit is the tiniest change an atom can experience. A quantum leap is actually a pretty insignificant thing!

We think of an electron orbiting a nucleus much like a planet orbiting a star, but this is just a convenient model. In quantum mechanics, the position of an electron is really more of a probability than a location.

# WHITE WATER

There's something quite mesmerizing about water flowing down a river, especially when there are obstructions such as rocks or constrictions caused by structures.

When this happens you might see a smooth, glassy surface as the river makes peaceful progress, until it suddenly accelerates and follows a dramatic downward curve. The smooth surface continues until it hits an area of boiling turbulence and chaotic white water, with eddies and vortices appearing and being swept downstream.

Kayakers and rafters seek out these places for the challenge they present, and artificial white water courses make for thrilling experiences for participants and spectators alike.

There is some fascinating physics going on beneath the surface, and the fluid dynamics demonstrated here are important in everything from making vehicles more aerodynamic, to checking that plumbing pipes are big enough.

## Go with the flow

The main external force on the water is gravity, drawing it downhill. The mass of the water has its own inertia, which resists any acceleration, and viscosity which makes it resist flow.

Viscosity is a measure of how "thick" a fluid is. Water is not very viscous compared to oil or honey, which means that it will travel more quickly in the same conditions.

The river upstream of our obstruction, where the water flows smoothly, is traveling relatively slowly. On reaching the obstruction, whether a submerged rock or weir, or something like a bridge pier, it has to accelerate. The channel in which it is flowing has become smaller, and for the same volume to pass it must do so more quickly.

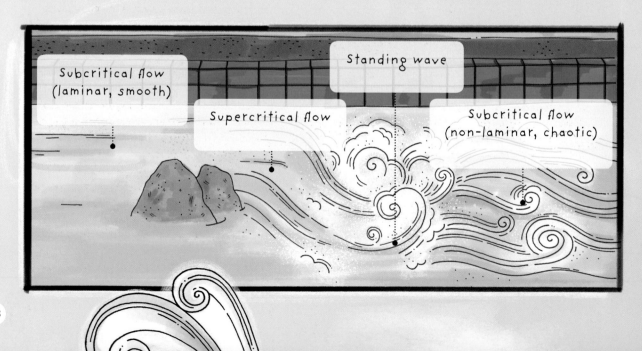

Subcritical flow (laminar, smooth)

Standing wave

Supercritical flow

Subcritical flow (non-laminar, chaotic)

## Fig. 1 – Water depth at subcritical and supercritical flow

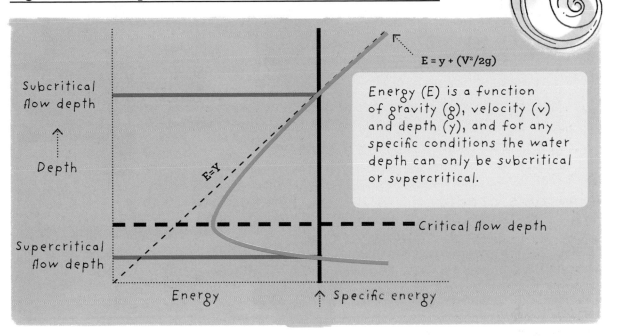

Subcritical flow depth

↑

Depth

E=Y

$E = y + (V^2/2g)$

Energy (E) is a function of gravity (g), velocity (v) and depth (y), and for any specific conditions the water depth can only be subcritical or supercritical.

Critical flow depth

Supercritical flow depth

Energy     ↑ Specific energy

# The shape of water

You might think that the water could make the opening bigger by raising its surface, so it doesn't need to accelerate – but in fact the water surface actually dips instead. This effectively makes the opening even smaller, so the water has to accelerate even more. That seems a bit, well, odd!

The scientific explanation is that the water is behaving in three quite distinct ways. Upstream, the smooth river is calm and has laminar flow, meaning the streamlines are (more or less) parallel. The viscosity of the water keeps everything tidy. This is subcritical flow, and is the normal condition.

The acceleration needed to pass the restriction increases the inertial forces on the water, and it then does something quite interesting – it changes its behaviour from subcritical (viscous forces dominating) to supercritical (inertial forces dominating).

Supercritical flow is much faster than subcritical, and is even faster than is needed to get the volume of water past the obstruction. This means that the free surface of the water actually drops, yet the full volume can still get past the restriction.

For any given flow conditions in a stream or channel, the water will be happy at either of two distinct depths, called the critical depths. One is subcritical, the other supercritical, and the only way to arrange for the water to remain at any other depth is to change the volume of flow.

Supercritical flow is when a fluid is traveling faster than the speed of a wave. Throw a pebble into a smoothly flowing stream and the circles of ripples will spread out in all directions evenly. In supercritical flow, the ripples are carried downstream faster than any can move upstream. This is broadly what happens when an aircraft goes supersonic (see Going rreeaallllyy fast, page 166).

Supercritical flow is laminar and smooth, but before long we encounter a third condition when the fast-flowing water suddenly boils into a turbulent mess as it encounters a stationary or standing wave. This marks a change back into subcritical flow, but now it is chaotic and non-laminar. The standing wave is also known as a hydraulic jump, because the water jumps upward. This is the best place to play in a raft or kayak, if you've got the skills.

# EXECUTIVE TOYS

Even the most powerful people like to fiddle with desktop gadgets from time to time. They might not appreciate it, but they are harnessing the principles of physics.

## Newton's cradle

The best known and possibly the original executive toy is Newton's cradle, even though Isaac himself never played with one and certainly didn't invent it.

A row of shiny steel balls are suspended from a frame so that they are just touching. When an end ball is raised and released, it strikes the row and stops immediately, while the farthest ball suddenly leaps into action. It then falls back, transmitting its energy once more through the row and making the first ball arc upward just as before.

Try lifting and releasing two balls, and both will stop dead while the two farthest away will arc upward. With three or four (in a five-ball cradle), the middle balls merrily swing back and forth while the end balls come along for the ride or wait motionless for the others to return.

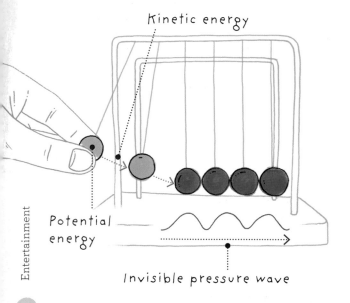

Kinetic energy

Potential energy

Invisible pressure wave

## Energy conversion at your fingertips

It's great fun, as well as being a cool demonstration of some basic principles of physics. What we've got here is a continuous conversion of energy from **kinetic** to **potential** and back again, along with a demonstration of the **conservation of momentum** and an invisible **pressure wave** running through the balls.

Lifting the first ball gives it potential energy, because gravity can now act on it. This becomes kinetic energy as it is released and accelerates. At the moment it hits the stationary ball, its kinetic energy drives a pressure wave causing the steel to compress and rebound slightly, and in doing so, transfers the wave to the next ball. It does this fast – at the speed of sound, which in stainless steel is about 5800m/s (that's nearly 13,000mph, or about 17 times faster than sound in air).

At the end of the row, the pressure wave's energy becomes kinetic energy again, kicking the last ball upward, exchanging kinetic for potential energy...and the process continues.

This can continue for some time, but not for ever. The energy in the whole system gradually reduces because of friction, air resistance and noise (every click you hear takes a little energy away), and eventually you are left with a set of balls swaying gently in the frame.

Most Newton's cradles have five suspended balls, but it works with any number (although it's a little boring if there aren't at least two).

# The fidget spinner

A short-lived craze that's more likely to be seen in the school playground than the executive desk, fidget spinners provide an intriguing feedback sensation through the fingers. This is caused by the gyroscopic effect, which responds to movement with a counterforce at 90 degrees.

Previous generations were similarly amused by toy gyroscopes, which perform improbable feats of balancing as long as they keep spinning. It's all to do with **angular momentum** and the spinning object's **moment of inertia.**

The heavy outer bearings on a fidget spinner give it a high moment of inertia – if they were closer to the centre, the moment would be lower. The high moment in turn means that angular momentum keeps the spinner spinning on the same axis, and resists attempts to change alignment.

Gyroscopes using the same principle are used in navigation instruments, because they always keep pointing in the same orientation.

How long can you keep your spinner spinning? My record is one minute fifty-seven.

## How to turn a bicycle

You think you know this, but you don't...
Pedal at a steady speed, and take your right hand off the bars.
Gently (gently!) pull the left bar toward you*.
Which way will you turn?

It might seriously freak you out that the bike leans and turns to the right, not the left. What you are experiencing is the result of the gyroscopic effect on the front wheel, which responds to your attempt to move the wheel's axle by reacting at 90 degrees. You pull left, it tips right.

*Seriously, make sure there's no traffic around before you try this.

Entertainment

# THE ELECTRIC GUITAR

Few inventions have done as much to change music, or create so many banging tunes, as the electric guitar. This discards the simple sound box of an acoustic instrument in favour of one or more sets of electronic pickups, and feeds a tiny electrical signal to an amplifier. Turn it up to 11. Understanding the physics that makes this possible won't necessarily improve your playing, but it can't hurt.

It's all down to that 19th century rock god, Michael Faraday. In 1831 he discovered that electricity and magnetism were intimately connected, and that any change in the magnetic environment around a coil of wire generates a voltage. Move the coil and the magnet closer, or apart, or rotate one with respect to the other, and an electrical signal is created.

This phenomenon of **electromagnetic induction** works both ways: move the coil or magnet to generate electricity, or add electricity to make the coil or magnet move. This basic fact is behind the workings of electric motors, transformers, dynamos, alternators, solenoids and...electric guitar pickups – the small electromagnetic sensors that convert string vibration into electrical signals.

## A new industry

Exactly a century later, in 1931, musician and inventor George Beauchamp formed a company called Ro-Pat-In to make instruments to exploit the Faraday law. His US patent, granted in 1937, described "a simple, practical and improved electrical musical instrument in which the vibrations of the sound-producing elements or strings directly vary the reluctance of a magnetic circuit to induce an electric current in a coil within the magnetic field, which current is suitably amplified and transformed into sounds as true reproductions".

Sadly, Beauchamp died in 1941 aged only 42 years, and so didn't live to hear Hendrix play "The Star-Spangled Banner", but his company (now Rickenbacker) became the first manufacturer of the new instrument.

A simple guitar pickup has a coil of thin wire wrapped around a small magnet, located below the metal guitar string. As the string vibrates it moves inside the magnetic field and so generates a small wave-pattern electrical signal in the wire. A high note vibrates more quickly, and a low one more slowly, in both the string and the signal.

Thinner strings vibrate more quickly, as do shorter strings (achieved by pushing the string against a fret). Thicker heavier strings produce the low notes.

This electrical signal can then be manipulated to distort the shape of the wave, for effects such as wah-wah or echo. It's then simply a matter of amplification to make sure they can hear you at the back.

Some guitars have a double pickup arrangement. The coils inside are wound in opposite directions, originally to reduce interference from stray electromagnetic signals that caused unwanted hum. These humbuckers also produce a rounder, more bass-like sound compared to the clearer note from a single coil pickup.

## Fig. 1 – Guitar pickup

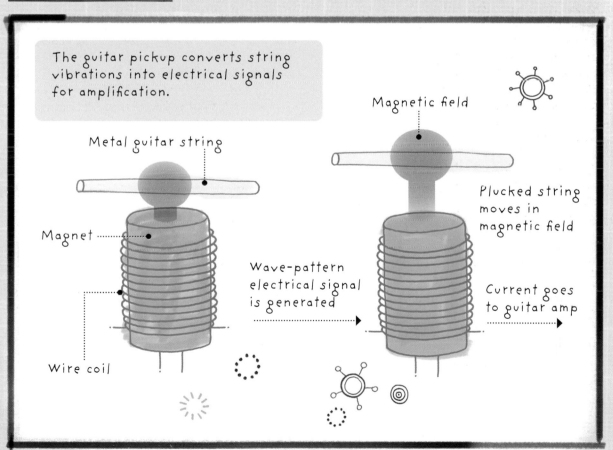

The guitar pickup converts string vibrations into electrical signals for amplification.

Magnetic field

Metal guitar string

Plucked string moves in magnetic field

Magnet

Wave-pattern electrical signal is generated

Current goes to guitar amp

Wire coil

# MUSIC

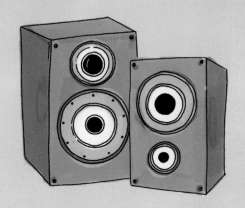

What's the difference between music and noise? If it is enjoyed by people more than about 20 years younger than you, it's probably noise.

**A** slightly more scientific answer is that music is an ordered structure of **compression waves** traveling through the atmosphere, while noise is an unstructured series of waves.

Compression waves are a bit like ocean waves in that they transmit energy without actually moving very much themselves. It's convenient to think of them as being wave-shaped, sinusoidal, like waves on the surface of water, but this can be a bit misleading. They are really more like fuzzy expanding bubbles of atmospheric compression and rarefaction.

The atmosphere, a mixture of gases, is quite easy to compress and also quite elastic – it wants to even itself out so that the pressure is the same everywhere.

When it is energized – for instance by the moving cone of a loudspeaker – the air molecules immediately in front are pushed forward and pulled back. These molecules then push and pull on the next front of molecules, which do the same in turn. Hey presto, a wavefront of higher and lower pressure emanates from the source of the sound.

This expands as it travels outward at the speed of sound – about 343 metres per second (767mph) – becoming weaker as its energy is spread over a larger area. The amplitude or volume falls according to the inverse square law, so at double the distance it will be just a quarter as loud.

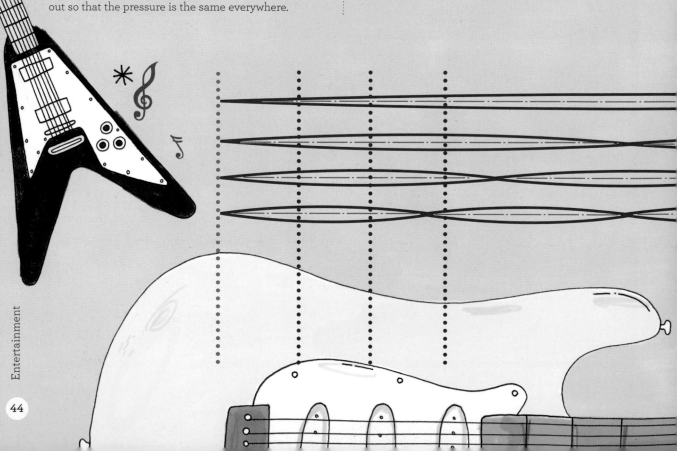

# The art of the sound of the noise of music

So much for sound, but what makes it music? The key aspects of sound wave functions – the physics of music – involve harmonic frequencies, resonance and overtones.

There is a huge variety of musical instruments, producing their sounds by being hit, plucked, blown or otherwise played. What they all have in common, every one of them, is that they vibrate the air somehow, sending compression waves into the atmosphere.

Every object in the Universe has a natural **resonant frequency**, which is the speed it will vibrate at when stimulated. Larger and heavier objects have lower frequencies, smaller and lighter ones, higher. It is not just the mass that is important, but also their configuration.

The strings on a guitar provide an example. The higher notes are produced by the thinner and lighter strings, and the note produced also depends on the tension in the string – its configuration.

The string is fixed at both ends, so these points must be nodes on any waveform the string adopts when plucked. A node is a point where a wave meets the baseline, and where the string is essentially stationary. This means that the longest wave (the lowest frequency) at which the string can vibrate is twice its length. This standing wave (because it isn't traveling) is called the string's fundamental frequency or first harmonic. The string can also vibrate so that a full wave exists between the two fixed nodes – the second harmonic. A wave and a half is the third, two full waves the fourth and so on. We hear the first harmonic most strongly, with the other notes – the overtones – adding depth and colour to the sound.

It's worth noting that guitar strings are thin, and when they vibrate they can easily slip through without displacing very much air at all. Unless amplified by electronics or the resonant cavity of the instrument's body, the sound they make is of very low amplitude.

In wind instruments, the situation is a little different and depends on whether the instrument is open at both ends (like a flute) or only at one end (such as a clarinet).

Exciting the air molecules sets up a standing wave in the bore. A closed end becomes a node in that wave, because the air cannot move past it. But an open end has no such restriction, and here the wave pressure will always stay at (more or less) atmospheric pressure.

This means that the fundamental wavelength when both ends are open will be twice the length of the pipe, similar to a guitar string. In a pipe closed at one end, the fundamental wavelength is twice as long again, because the open end is a peak or trough in the waveform. In other words, a clarinet will produce a fundamental note at half the frequency of a flute with identical internal dimensions.

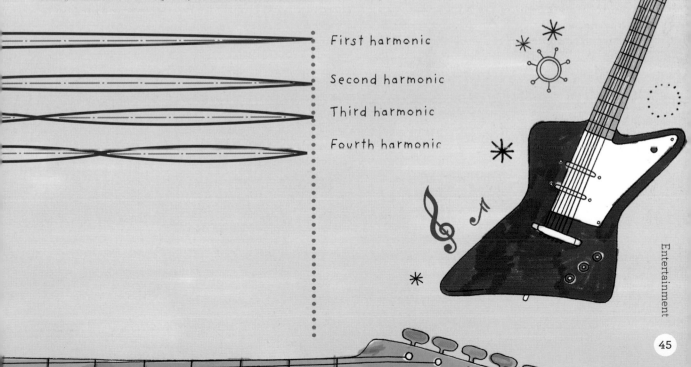

First harmonic

Second harmonic

Third harmonic

Fourth harmonic

# Sonic beam

A feature of sound is that it usually propagates in all directions, and so is detectable over a large area. This also means that its energy diminishes rapidly with distance.

A clever new method makes it possible to create a tightly-focused beam of audio energy, which can be directed at a particular spot. Anybody in the spot, which might be hundreds of metres from the audio source, can hear the sound perfectly. Anybody just outside it hears nothing.

The trick is the use of high-frequency ultrasound, and the ability of waves to combine and interfere constructively or destructively.

Human hearing ranges from about 20Hz to 20kHz. Direction audio devices work at frequencies around 200kHz, which is completely inaudible to humans or dogs. Two waves are generated – one a steady reference tone at 200kHz, the other a modulated wave at 200.2 to 220kHz.

The modulated wave carries the sound signal being transmitted, but shifted far up the audio spectrum.

High-frequency sounds have short wavelengths, which refract less than longer ones. What this means is that the ultrasound can be focused, like an invisible sonic searchlight.

200kHz reference wave

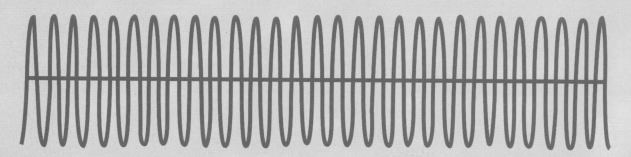

**+**

Audible sound

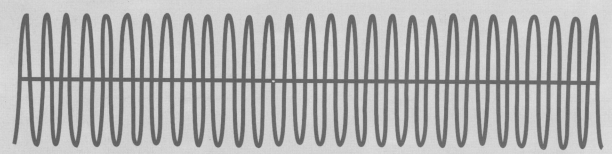

The two ultrasound signals are beamed together and remain inaudible until they reach a solid object such as a person. Then the magic happens: the beams are abruptly slowed down, and combine their energy. Where wave peaks and wave troughs match, the result is **constructive interference**; where peak meets trough the result is **destructive interference**.

The 200kHz reference wave effectively decodes the modulated wave back into the human hearing range, and audible sound is produced at that point. It's a spooky experience, and surprisingly effective.

Directional audio can be used at sea to warn specific vessels of dangers without deafening an entire ocean, or to send a message to a person in a crowd. It might even provide a hands-free speaker phone where only you can hear the caller.

The two waves combine, decoding the original <u>message</u> <u>back into human hearing</u>.

=

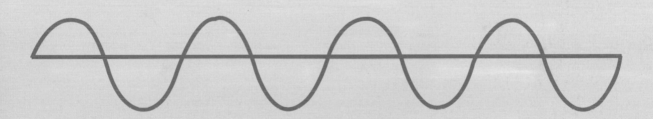

# 3D MOVIES

Everybody loves a good movie, even if not everybody agrees which movies are good. A powerful story, convincing acting, maybe even top-notch special effects can all build up an illusion of being somewhere else, watching something that seems quite real.

I f the screen is large enough to fill most of your field of vision, and there are few distractions, the effect can be almost three-dimensional even on a standard two-dimensional screen. Sometimes, though, a bit of clever physics can generate a 3D effect from a flat screen, and then the illusion becomes even more convincing.

All of the techniques work by fooling the brain into believing that the image has depth, by delivering slightly different pictures to each eye. The pupils in your eyes are about 6cm (2.5 inches) apart, so when viewing a real scene they each get slightly different information. The amazing processing power inside your head converts this into depth perception.

To appreciate how important vision is, consider the fact that our brains are relatively huge compared to other animals, yet they still devote about a quarter of their computing power to image processing.

That also means that we have to try quite hard to fool ourselves into seeing 3D.

The simplest method is the **anaglyph** technique. Practically all real-world colours can be reproduced with varying strengths of the three primary colours – red, green and blue. A reasonable approximation reduces this to just two – red, and a combination blue/green called cyan.

If a scene is filmed through two lenses, about 6cm (2.5 inches) apart, with a red filter over one lens and a cyan over the other, the resulting image exhibits strange coloured haloes – most noticeable around objects nearest to the camera. View the image again, with a red filter over your left eye and a cyan over your right, and the image pops into 3D.

## Tricking the eyes to deceive the brain

A different approach makes use of the peculiar qualities of **polarized** light. You may be familiar with the way polarizing sunglasses reduce reflections from flat surfaces such as lakes and snow. Normal or "incident" light vibrates in all directions – up and down, left and right and all angles in between. Being reflected from a horizontal surface changes these vibrations, so that about half of the light now vibrates left–right. The light has become polarized. Polarized sunglasses are marked with microscopic parallel lines running up and down,

so close together that light vibrating left–right cannot get through. It's like trying to get an envelope though a letterbox the wrong way.

This effect can be exploited in a way similar to the anaglyph technique. The scene is filmed through two lenses, with polarizing filters at 90 degrees to each other. When later shown together, the information destined for each eye is unscrambled by a pair of viewing glasses with similarly arranged filters.

Both of these **passive 3D** methods are relatively easy and cheap, using simple unpowered headsets. **Active 3D** systems are said to offer better quality and fewer headaches, but are complex and require powered glasses. These incorporate liquid crystal shutters that alternately allow each eye to see the screen, and mean that careful synchronizing of the images with the glasses is necessary.

Can we achieve 3D without the cumbersome glasses? This is possible, but it only really works on small screens such as a phone or games console. Here, a carefully engineered **parallax filter** can send different images to each eye – the eyes are quite close to the screen anyway, and so already see it slightly differently. The effect is similar to the 3D printed images with a ridged plastic surface (lenticular 3D) that are sometimes used on greeting cards or posters.

## Fig. 1 – Sending different images to each eye

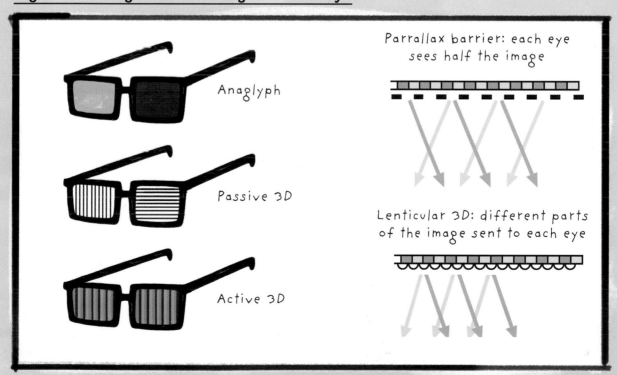

Anaglyph

Passive 3D

Active 3D

Parrallax barrier: each eye sees half the image

Lenticular 3D: different parts of the image sent to each eye

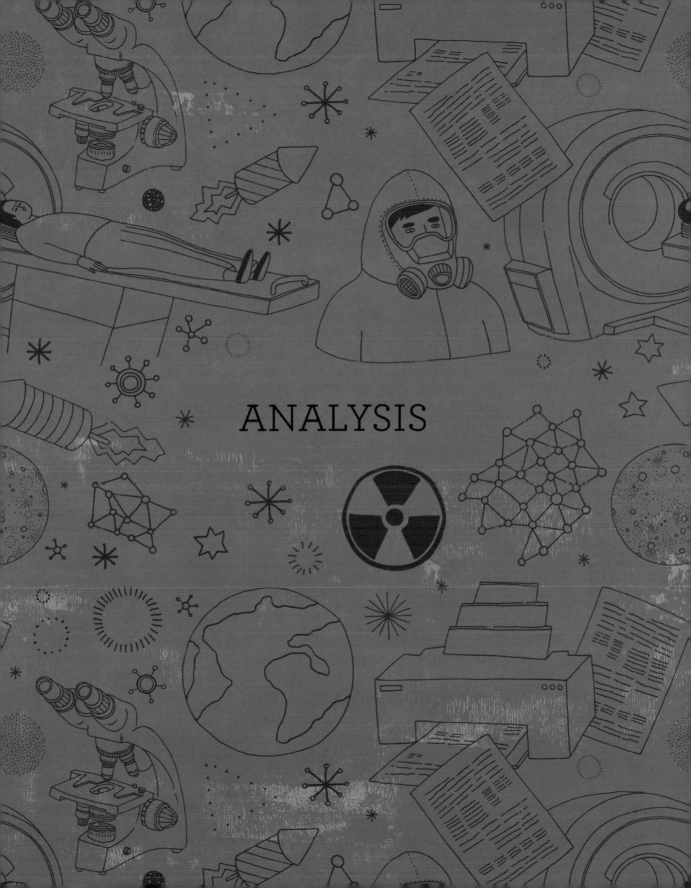

ANALYSIS

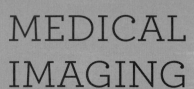

# MEDICAL IMAGING

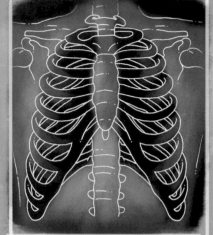

If you're having an X-ray on any part of your body, chances are you'll be more concerned about what it might show than the way it shows it. X-ray images are made by photons just like those in visible light, but with much shorter wavelengths and consequently much higher energy. It is this high energy that allows them to pass through human tissues, producing an image of what is inside the body.

The startling ability of **X-rays** to pass through many objects was discovered in 1895 by Wilhelm Röntgen, in one of science's famous accidents. Experimenting with a discharge tube – a glass bulb in which electricity can be passed through gas – he noticed a screen in the darkened room glowing even when the tube was covered. He stepped in front of the screen and was astonished to see a projection of his own skeleton.

Since then medical X-rays have come a long way, with the original wet photographic process being replaced by digital imaging similar to digital cameras. X-rays are used in CT scans, for baggage security at airports and in industrial quality control.

What Röntgen didn't realize, when cheerfully demonstrating his mystery rays (he called them X-rays because he didn't know what they were), was the risk to health that comes from ionizing radiation like this. The penetrating power that allows X-rays to pass through objects also means that they can damage cells and cut strands of DNA, possibly causing long-term health problems. This can be turned to advantage by targeting cancerous cells with X-radiation.

## Shades of grey

The key to the usefulness of **radiographic imaging** is the variable transparency of tissues and bones, depending on how dense they are. Radiographers work on the basis that there are only five shades in an image, but experienced eyes can discern detail by finding edges of similar looking greys.

Hollow spaces like healthy lungs appear almost black, while fat appears dark grey. This is evidence that fat is actually less dense than muscle, which appears as mid-grey. Bone is revealed as light grey, while anything that appears white is probably metal, perhaps an implant or jewellery.

This grey scale is also used for CT scans, which are essentially a 3D version of X-ray images. Instead of simply grabbing a single image of a patient, a **computerized tomography** (CT, or CAT) uses a number of beams of X-rays and several detectors at once. The rays are usually fired from a ring-shaped device, with the patient inside the ring, and may move back and forward and rotate during the exposure.

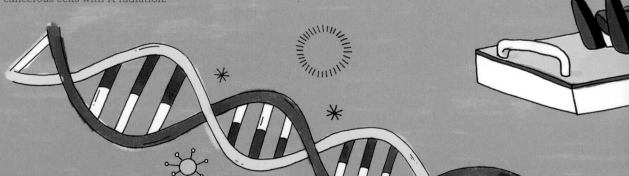

# MRI

Although X-ray and CT scans can make images of soft tissues, better detail can be obtained using **magnetic resonance imaging (MRI)**. This exploits the properties of the water molecules that make up about 70 percent of the human body.

A strong magnetic field makes the hydrogen nuclei (which are just protons) in water line up parallel to the magnetic field lines. Most adopt a low-energy state along the lines, but some take a high-energy alignment facing the opposite direction. They are then bathed in radio energy, causing some of the low-energy protons to flip into the high-energy state. When the radio pulse ends, the protons return to their previous alignment and release some energy, which is picked up by radio coils around the machine.

Different types of tissue relax at different speeds, and this information is used by the computer to create detailed images showing different tissues with high levels of detail.

Modern MRI scanners have magnetic flux density strengths of up to three tesla, which is about sixty thousand times as powerful as the Earth's magnetic field.

There's no radiation dose with MRI, but you may not be able to have one if you have metal implants.

CT is an advanced X-ray technique that takes many virtual slices through the body parts being examined, useful for complicated fractures and investigating conditions with different tissue densities, such as tumours.

MRI can reveal greater detail in soft tissues, for instance with sports injuries.

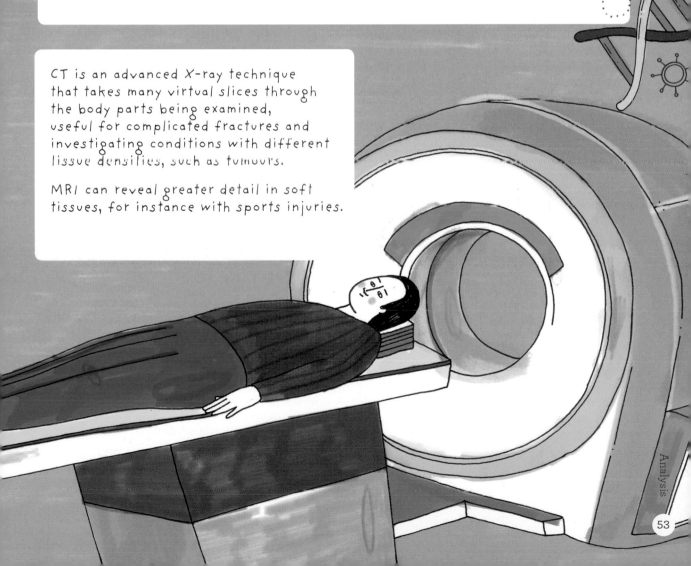

# Radiation doses

We're all exposed to radiation every day, and this is nearly always harmless. Outside, we feel the warmth of the Sun through infrared radiation, and see via light. These low-energy wavelengths are of no concern. Ultraviolet (UV) is higher energy, causing tanning and potentially skin cancer. There are X-rays and gamma rays reaching us from space, but the atmosphere blocks most of them. There is also background radiation from the soil, from nuclear tests and accidents, and from a variety of other sources.

Ionizing radiation doses are measured in **sieverts**, describing the energy absorbed per unit mass. In Taiwan the average person gets an annual dose of about 1.56mSv, in the UK it is 2.7 millisieverts (mSv), and in the USA and Finland it is more than double this at 6.2mSv and 7mSv, respectively.

Why the difference? Mostly this is down to a higher number of X-ray examinations carried out, although other factors such as frequency of flying and the amount of radon gas in the environment are also involved.

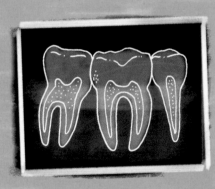

### 0.014mSv
### Chest X-ray

### 0.01mSv
### 100g Brazil nuts

If you ate 500g of Brazil nuts a day, every day for a year, you would reach the exposure limit for nuclear workers – but that would be the least of your problems.

### 0.08mSv
### Transatlantic flight

A single transatlantic flight gives a dose of 0.08mSv. If you were ever lucky enough to fly on Concorde, you might think you had cut your exposure by half due to the faster crossing, but the higher altitude almost exactly canceled this out. Various studies have shown that flight crew get about the same radiation dose from flying as from everyday life.

### 0.005mSv
### Dental X-ray

A dental X-ray gives a dose of about 0.005mSv, while a chest X-ray is about 0.014mSv. A CT scan delivers many times this, as high as 10mSv for a full spinal scan.

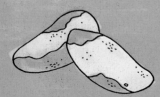

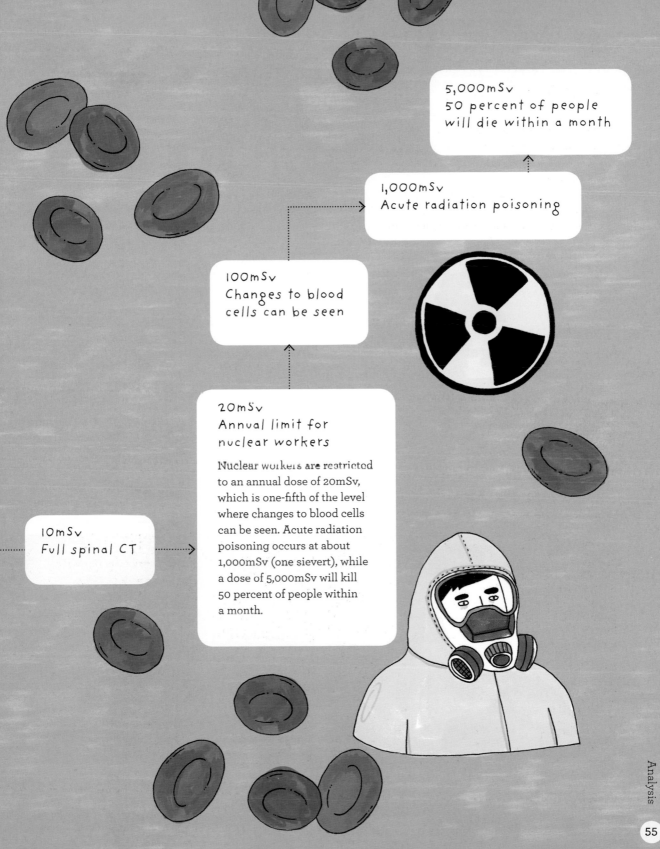

5,000mSv
50 percent of people
will die within a month

1,000mSv
Acute radiation poisoning

100mSv
Changes to blood
cells can be seen

20mSv
Annual limit for
nuclear workers

Nuclear workers are restricted
to an annual dose of 20mSv,
which is one-fifth of the level
where changes to blood cells
can be seen. Acute radiation
poisoning occurs at about
1,000mSv (one sievert), while
a dose of 5,000mSv will kill
50 percent of people within
a month.

10mSv
Full spinal CT

# LOOKING AT LITTLE THINGS

What's the smallest thing you can see? A grain of sand might be a millimetre across, and a human hair about a tenth of that. Anything smaller is difficult to see properly with the naked eye – you might catch sight of a speck of dust in the air, perhaps a hundredth of a millimetre across, but you can't actually see the detail of it.

For that you need a **microscope**. A good quality optical (or light) microscope reveals a world of very tiny things invisible to the naked eye, but all instruments have their limits. At the extreme, the resolution of a light microscope – the smallest distance between two points that can be seen – becomes a function of the wavelength of light. Anything less than about half a wavelength becomes impossible to see.

Visible light has wavelengths between 400 and 700nm, meaning white light has an average wavelength of 550nm. The resolution limit for light microscopes is therefore about 250nm.

To see small objects with detail that light microscopes never can, we need a different tool. **Electron microscopes** use (as the name suggests) electrons instead of photons (particles of light). Electrons vibrate with a wavelength of around 1nm (this is the realm of quantum mechanics, and things quickly get complicated), enabling us to see hundreds of times better than with visible light.

In an electron microscope, the lenses are electromagnetic coils. A stream of electrons is produced by heating a wire or crystal, and the electrons are accelerated to strike the specimen. The whole process takes place in a high vacuum.

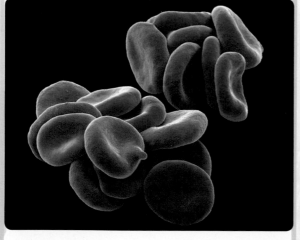

Blood cells

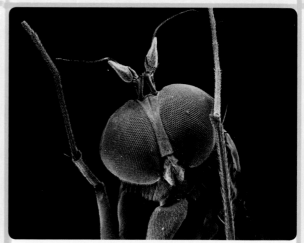

Fly

# Electron microscopes

Not all electron microscopes are the same. In **transmission electron microscopy (TEM)**, electrons are focused onto very thin samples, causing secondary electrons to be driven off the other side. Here they can be collected on a sensor or even a screen and eyepiece to create a visible image. Preparing specimens for TEM is a skilled business, as samples may need to be sliced to just 50 or 100nm – a thousandth the thickness of a human hair.

The other main type is **scanning electron microscopy (SEM)**, in which a tight beam of electrons is drawn across the specimen line by line. Electrons strike the sample and release secondary electrons, which are recorded as a signal. A computer then assembles the information into a detailed image. Unlike TEM, SEM images show depth.

A third technique, **scanning transmission electron microscopy (STEM)**, combines elements of both other SEM images and are often published with beautiful pastel colours to highlight the objects revealed, but these do not represent any true colour. Features may be smaller than wavelengths of light, so it can be argued that they cannot actually have any colour at this scale.

Both SEM and TEM samples are coated with a vanishingly thin layer of metal, usually gold or gold-palladium, to make them electrically conductive and prevent stray electrons from damaging the images.

This means that when you see an electron microscope image, what you are actually looking at is a false-coloured Photoshopped® image of the atoms-thick gold coating sprayed onto the sample.

Tardigrade

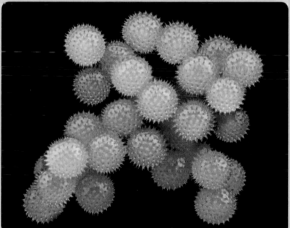

Pollen grains

# SPECTROSCOPY

It might sound like some super-complicated laboratory technology that only a genius with multiple degrees could understand, but the science behind spectroscopy is both surprisingly simple and endlessly fascinating. This is a method to discover what something is made of just by checking its colour.

We might think of the colour of an object as, say, red or yellow, or use a reference in a paint catalogue, or a Pantone® number. A colour on your phone or computer screen is described by the strength of three primary colours – red, green and blue, to give RGB values. The colours in this book are made with four inks coloured cyan (bluish), magenta (reddish), yellow and black – CMYK.

Both RGB and CMYK show that what looks like a single colour is actually made up of several components. **Spectroscopy** – the analysis of the spectrum – uses the same approach.

Isaac Newton realized in 1665 that sunlight could be split into a spectrum of colours by passing it through a glass prism, and (crucially) that those colours could be recombined to make white light. This was the first spectroscopy experiment, although it took another 200 years before Gustav Kirchhoff and Robert Bunsen (he of Bunsen burner fame) made the first proper spectroscope instrument to study the components of light.

The device allowed them to compare the spectra from different light sources, such as flames, and see that some were stronger in certain wavelengths and weaker or absent in others. The breakthrough came when it was realized that every chemical element produced a unique pattern of bright spectral lines when a sample was burned.

At the atomic level, burning an element causes its electrons to jump to a new energy level and, in doing so, release photons of light. The wavelength of these photons is unique to the element, because the pattern of electrons in the atom is also unique.

This is what gives fireworks their wonderful range of colours. If a pyrotechnic is loaded with strontium, the resulting flash will be red; copper gives blue, and sodium yellow. Each time a firework explodes, you see the photons released because of the quantum jumps performed by the electrons orbiting the atoms of the element being ignited.

# Elemental colours

Back on the ground, we can use this knowledge of the relationship between colours and elements to find out lots of useful information. When the light of the Sun or a distant star is analysed, its **continuous spectrum** shows a series of dark lines. These absorption lines reveal what elements exist in the star's upper layers, or in the space between the star and Earth. This is because cold elements absorb the same wavelengths that they emit when heated.

Spectroscopy is also one of the most powerful techniques for non-destructive analysis of materials, particularly using near-infrared (NIR) spectroscopy. In this technique a beam of infrared light is shone at a sample, and the reflected wavelengths analysed.

It is a fast and simple technique, but only reveals information about the surface of a target.

Three different types of spectrum are analysed. A hot opaque object such as a lamp or star produces a continuous spectrum over a range of wavelengths. When the light from a hot object passes through a transparent medium (the glass in a lightbulb, or interstellar space) some wavelengths are absorbed and the result is an **absorption spectrum**. If that medium were to be heated, the same black absorption lines would appear as bright lines in an **emission spectrum**.

Each of these is like a chemical fingerprint of the elements that make up the object, revealed by their colours.

## Fig. 1 – Spectrum analysis

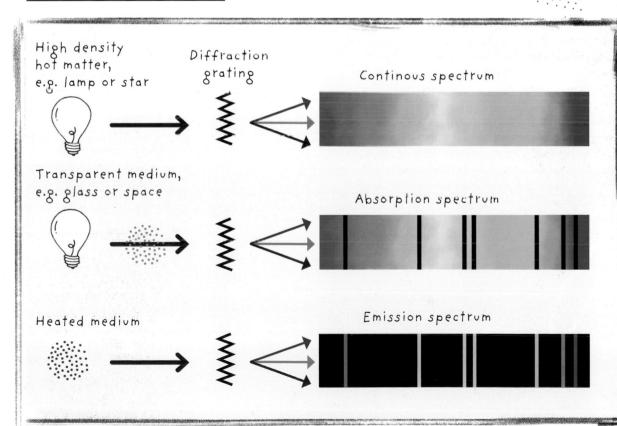

High density hot matter, e.g. lamp or star

Diffraction grating

Continous spectrum

Transparent medium, e.g. glass or space

Absorption spectrum

Heated medium

Emission spectrum

# CRIME SCENE INVESTIGATION

The world of forensic science has been revolutionised by the physics of mass spectrometry, one of the most powerful tools for analysing evidence left behind at the scene. But what is a mass spectrum, and how does it reveal who might have done it?

A mass spectrum is a graph representing the concentrations of individual elements in a sample being investigated. The elements are sorted by mass, and aggregated to show how much there is. **Mass spectrometry** might be most familiar from TV forensic investigation dramas, but it is a powerful tool in real world scientific investigation. Don't confuse it with the similar-sounding spectroscopy (see page 58), which is a kind of advanced colour analysis.

To break down a sample into its constituent elements and isotopes, it is placed in a vacuum chamber, heated and bombarded with electrons. This causes its atoms to become ionized, or electrically charged, by having electrons added or knocked off.

The ionized atoms are then accelerated inside the instrument, first through an electrical field between two highly charged plates. The more highly ionized particles accelerate most, beginning the sorting process. The particle beam now enters a strong magnetic field, which applies a turning force, which sorts the particles by mass and charge. Lighter components are turned most, due to their lower inertia, while the heavier ions deflect least. Strictly speaking, they are sorted by mass/charge ratio rather than simply mass.

Particles are recorded as they strike an ion detector at the end of the tube. There may be an array of detectors to record different masses simultaneously, or a single detector which is moved through the particle beam to record impacts. Either way, the signals build into the mass spectrum indicating the relative abundance of each component in the sample. Comparison with reference spectra then provides clear evidence of the chemical make-up of the sample.

A slightly different sorting technique is called time-of-flight (TOF) mass spectrometry, in which the ions are given a magnetic kick and the time taken for them to reach a detector is recorded.

## Gas chromatography

Mass spectrometry is often coupled with a gas chromatography, in which the constituent components of a mixed sample are "unmixed" by being passed through a separation column. This combined method, GC-MS, enables complex samples to be first broken into individual substances and those substances identified by their mass spectrum.

If you watch a lot of crime scene drama, you'll begin to recognize some of the instruments. You might even see the MS being applied before the GC, which is guaranteed to horrify any actual forensic scientists watching.

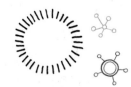

## Fig. 1 – The mass spectrum of caffeine

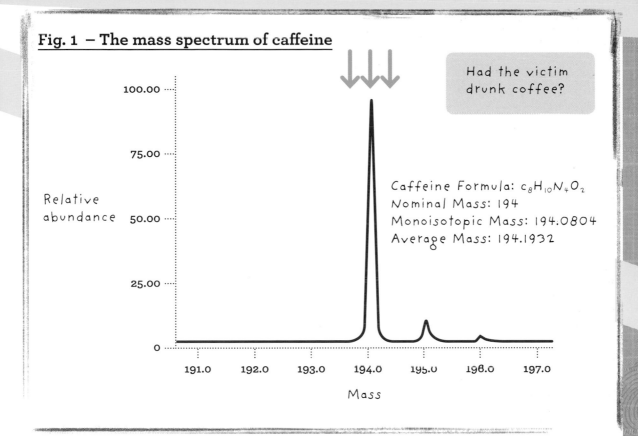

Had the victim drunk coffee?

Caffeine Formula: $C_8H_{10}N_4O_2$
Nominal Mass: 194
Monoisotopic Mass: 194.0804
Average Mass: 194.1932

## Mass spectrometry on Mars

The main science payload on the Curiosity rover, presently trundling around the surface of Mars, is a compact laboratory including a gas chromatograph, mass spectrometer and laser spectrometer. In typically cutesy NASA fashion, it is called Sam – "sample analysis at Mars". This 40kg (88lb) package makes up half the total mass of scientific instruments in the 900kg (2,000lb) vehicle.

This mass spectrometer is of the quadrupole type, in which ions are accelerated and filtered between four conductors energized with both AC and DC current. It's a compact and robust instrument for this challenging away mission, and its primary purpose is the search for organic compounds that might help scientists on Earth determine whether life has ever existed on the red planet.

# TRICORDER

There might be something in the assertion that life imitates art more than art imitates life, at least as far as science and engineering are concerned. The last few decades have seen an explosion of new technologies and devices, quite a few of them looking like they might have been inspired by science fiction classics – especially *Star Trek*.

**W**e don't yet have warp drive or transporters, but we do have communicators (mobile phones) and replicators (3D printers).

Another iconic technology of Starfleet is the **tricorder**, an all-purpose portable sensing instrument that could diagnose illness and determine the composition of materials, without making any actual contact with them. Wouldn't it be cool to have something like that in real life?

This could well become the next example of life imitating art. **Non-contact sensing** is already here in both medical and non-medical applications, and many of the features of the sci-fi instrument are available today. Non-contact spectroscopy is routinely performed via handheld near-infrared (NIR) instruments (see Spectroscopy, page 58), and it isn't too hard to imagine that wide-spectrum analysis could become available in a similar way in the future.

What about medical tests? Analysis of exhaled breath is one of the more interesting frontiers in biomedical science, and it is already known that organic compounds called ketones can provide an early indication of diabetes – even before any actual symptoms are experienced. Other volatile organic compounds (VOCs) can indicate some forms of cancer, and conditions including obesity and allergies also leave characteristic traces in the breath and body aroma.

## Something in the air

VOCs are routinely measured in the atmosphere using **photoionization detection**. A small probe sniffs the air, and a strong ultraviolet lamp ionizes its molecules. Detectors measure the electrical signal produced, which reveals the concentration of volatile organics in the air. This is sensitive to about one part per billion.

A completely different approach uses skin emissivity as a diagnostic tool. Emissivity is a property of materials that reveals how well they emit energy. Two similar-looking surfaces under identical conditions may have very different emissivity. A recent development in bioelectromagnetics measures skin emissivity in the millimetre-band wavelength (between infrared and microwaves) for non-contact diagnosis of a range of skin conditions.

Millimetre waves are also used in the latest airport security scanners, as they are capable of "seeing" through clothing to reveal concealed objects. Some of those objects might be part of a person's anatomy, making these scanners a bit controversial.

With a combination of NIR spectroscopy, atmospheric gas detection and millimetre wave scanning, we're well on the way to a handheld multi-purpose sensing instrument. All we need is to add a few more sensing technologies and combine it all into one portable device.

That, and a starship to carry it on.

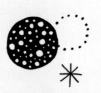

# The real tricorder

Such is the interest in making sci-fi into reality that the prestigious XPrize launched a competition in 2011 to develop a real working tricorder. The Qualcomm Tricorder XPrize of $2.5 million was awarded in 2017, not to a single device, but to a range of instruments collectively known as DxtER, created by a team called Final Frontier.

These are designed for consumer use, and can diagnose 34 different medical conditions including anaemia, tuberculosis and hepatitis A. Tests are non-invasive but not non-contact, and the device relies on a smartphone.

It's not quite something McCoy would understand as a tricorder, but then this is only the 21st century, not the 23rd.

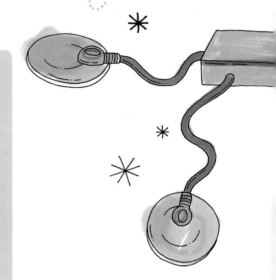

## Fig. 1 – Photoionization detector

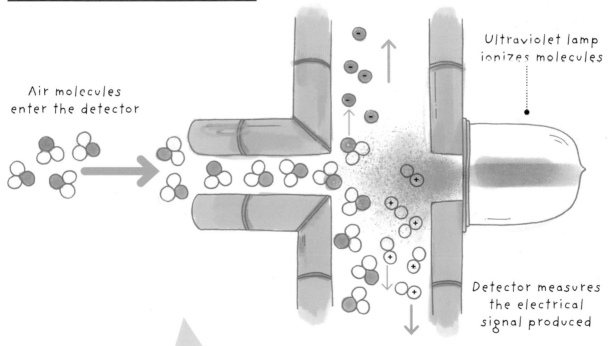

Air molecules enter the detector

Ultraviolet lamp ionizes molecules

Detector measures the electrical signal produced

# MICROFLUIDICS

If you have a computer printer, chances are that it will be of the inkjet type. These devices are quite simple and cheap to make, because a lot of the complicated precision engineering is located in the disposable ink cartridge (which is why these are so expensive). Inkjet printers are the most everyday example of a relatively new physical science that could have a big impact on the future. This is microfluidics which, as the name suggests, is all about manipulating very small amounts of liquid.

**M**icrofluidic devices have networks of tiny channels and other components created inside a solid material, usually plastic or glass. Channels can be as narrow as a few hundred nanometres or as wide as half a millimetre. The smallest might be just one-fiftieth of the diameter of a human hair.

Other components built into microfluidic chips include chambers where fluids can mix and react, pumps, valves, filters and detectors – all at millimetre or sub-millimetre scale. These do the same job in a microfluidic circuit as they might in a full-sized laboratory, processing plant or factory – which is why **microfluidics** is often referred to as lab-on-a-chip.

The comparison with electronics is inescapable. Where once electronic devices were heavy, relying on vacuum tubes (valves) and other large power-hungry components, today transistors and diodes are microscopic and incorporated into ever-tinier silicon chips.

## Miniaturized plumbing

The physics of fluids at tiny scales is different than at normal (macro) scales. The surface tension of a liquid is what causes a raindrop to form into a sphere, and a curved surface – the meniscus – to appear on a glass of water. Surface tension is an example of a viscous effect, and at microfluidic scales these easily outweigh the inertial effects that usually dominate fluid behaviour at the macro scale.

In practice, this means that **microfluidic flow** is laminar rather than turbulent (see White water, page 38), and so mixing of fluids is much harder to achieve. The technical explanation is that microfluidic operations take place at a very low **Reynolds number**. While this makes some operations more difficult, it can be an advantage when separating compounds for analysis. This is because, unlike at macro scale, the components do not immediately try to recombine once separated. With fluid volumes measured in femtolitres – quadrillionths of a litre – reactions take place very quickly.

The small scale of microfluidic processes also means that the thermal effects of exothermic (heat-generating) and endothermic (heat-absorbing) reactions can be easily dissipated. This can be a significant problem in industrial processes, but not at these scales.

## Microfluidics in everyday life

Microfluidic instruments are becoming commonplace in some medical applications, where they can provide laboratory-like analysis of body fluids at the bedside. This is much faster than sending samples away, and with mass-production of disposable sensors it can also reduce healthcare costs. The technology is also being exploited in some battery and fuel cell systems, which might allow laptops and phones to run for days or weeks between recharges, and could make electric cars as fast to refuel as internal combustion vehicles.

One potential application for microfluidics is in production rather than analysis, using the physics of small-scale fluids to extract and combine molecules in order to synthesize microscopic quantities of valuable compounds.

Sadly, none of this is likely to make your printer refill any cheaper.

## Reynolds number

The nature of fluid flow is revealed by the Reynolds number, a dimensionless quantity defined as the ratio of viscous to inertial forces within a fluid. Everyday experience of water is at a high Reynolds number and is of little use in microfluidics design because fluids behave differently at this scale.

## Fig. 1 – A simple lab-on-a-chip design

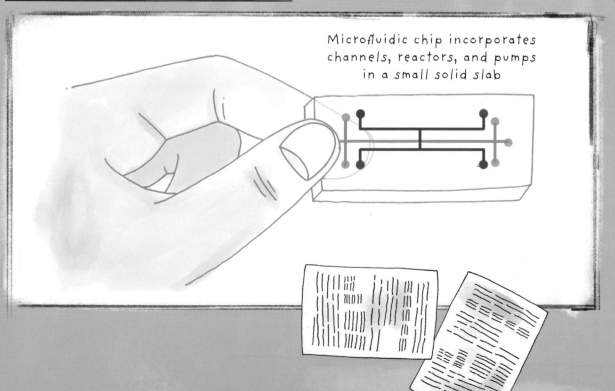

Microfluidic chip incorporates channels, reactors, and pumps in a small solid slab

# RADIOCARBON DATING

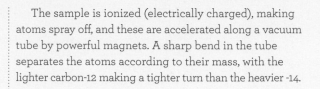

It's not unusual to hear of some archaeological discovery going back thousand or tens of thousands of years, and which gets scientists so excited that newspapers claim the textbooks will have to be re-written. But if a tooth or piece of wood is dug out of a bog or cave, how does anybody know how old it really is?

For any artefact that was ever part of a living thing, the answer is usually **radiocarbon dating**. This makes use of the phenomenon of radioactive decay, comparing the concentration of different isotopes of carbon in the sample. Carbon-12, carbon-13 and carbon-14 have very similar chemical properties but different masses because there are different numbers of neutrons in the atomic nucleus.

The heaviest, carbon-14, is naturally radioactive. Because of this, it experiences radioactive decay at a predictable rate, with the atoms gradually turning into nitrogen-14. That rate is the half-life of carbon-14, which has been measured at 5,730 years.

What this means is that every 5,730 years, half of the carbon-14 atoms will have spontaneously decayed to nitrogen-14. If you start off with 800 atoms, after 1 half-life you will have 400 left. Then 5,730 years later you will have half of the 400, or 200 and so on.

This is an incredibly useful technique for calculating the age of once-living objects. Carbon-14 is manufactured in the atmosphere by the action of cosmic rays, and the rate of formation seems to have stayed about the same throughout Earth's history. Any living object will have the same ratio of carbon-14 to carbon-12 as the atmosphere does, until it dies. At that point its carbon is locked up, and the -14 isotope quietly decays until some modern archaeologist comes along and starts interfering with it.

In a typical modern carbon dating set-up, an accelerator mass spectrometer is used to count the individual atoms of carbon-12 and -14 in a sample.

The sample is ionized (electrically charged), making atoms spray off, and these are accelerated along a vacuum tube by powerful magnets. A sharp bend in the tube separates the atoms according to their mass, with the lighter carbon-12 making a tighter turn than the heavier -14.

## The dating game

The result is a mass spectrum (see page 60 for spectrometry) – sorting the atoms into groups according to their mass. Atoms strike an ion counter, each impact is counted, and the rest is straightforward mathematics.

Carbon dating assumes that the rate of carbon-14 formation in the atmosphere has remained steady through time, and that the sample's carbon content is unaltered. By checking results against known dates – remains from ancient Egyptian tombs, tree ring calendars (dendrochronology) and sediments from long-lasting lakes, calibration curves have been developed which help improve accuracy. The technique is most accurate for younger samples, and is generally considered reliable up to at least 50,000 years.

In a single gram of carbon today, about 15 atoms of carbon-14 will undergo radioactive decay every minute.

## Fig. 1 – Carbon dating

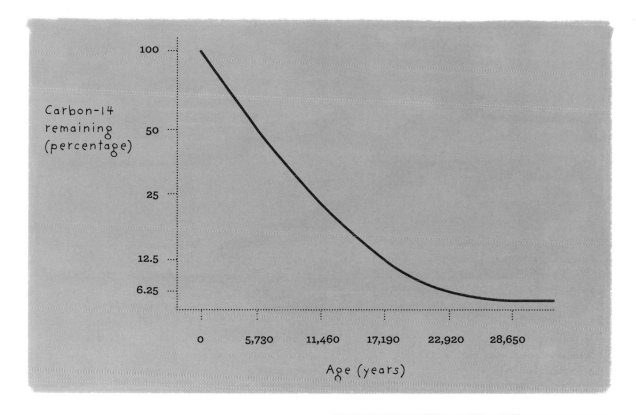

Carbon-14 remaining (percentage)

100
50
25
12.5
6.25

0    5,730    11,460    17,190    22,920    28,650

Age (years)

## Future archaeology

It's handy for us that we can use radiocarbon dating, but future archaeologists won't be so lucky. The ratio of carbon isotopes in the atmosphere is no longer just because of cosmic rays striking nitrogen molecules, but also from the burning of fossil fuels and nuclear tests.

Fossil fuels add carbon-12 to the atmosphere, but not carbon-14. Nuclear explosions (bombs and tests) have the opposite effect, increasing carbon-14 content. This means that a new and cleverer calibration will be needed by those scientists yet to come.

## Clocks in rocks

What about dating things that have never lived? Things such as the Earth or Moon?

The predictable tick-tock of radioactive decay is useful here too. Potassium-40 has a half-life of 1.25 billion years (and decays to argon-40), uranium-235 decays to lead-207 with a half-life of 704 million years and rubidium-87 becomes strontium-87 over a half-life of 48.8 billion years.

If you have a reasonable idea of the starting ratios of these elements, you can estimate the age of the rocks they are found in. The oldest rocks on Earth have been dated to about 4.4 billion years, while the oldest moon rocks may be slightly older. Meteorites have been dated this way to around 4.6 billion years.

# PROVING THE EARTH IS ROUND

There are people who, even today, say they think the Earth is flat. Some of them might just be saying that, but some do seem to believe it. This would be seen as hilarious by many of our ancient ancestors, who not only realized that the Earth is roughly spherical but had also made pretty accurate measurements of its size.

As early as 240BC the Greek scientist (and the inventor of geography) Eratosthenes used some fairly basic geometry to measure the circumference of the planet. If he could do it, so can you. What's more, with some simple observations and common sense, you can measure the distance to the Moon and the Sun. It will help if you can enlist a friend, preferably in another city roughly north or south of you.

Eratosthenes got a clue to the method when he discovered a well in the desert in Egypt where something remarkable happened each summer solstice (21 June on modern calendars). At noon, the Sun shone straight down the well and illuminated the water at the bottom.

## Measuring shadows

The well was vertical, and so the Sun must be directly overhead. Where he lived in Alexandria, some distance north, a tall vertical tower cast a shadow all day. Therefore the Sun was not directly overhead in both places, and indeed there was never a day when it was directly over Alexandria.

By measuring the length of the shadow at Alexandria, he calculated that the Sun was a one-fiftieth of a full circle away from overhead (we'd call this about 7.2° in modern geometry). With his logical mind he realized that the distance between Alexandria and the well (at Syene, today known as Aswan) must therefore be one fiftieth of the circumference of the Earth.

He worked out a distance between the two, did his sums, and got an answer that was within about ten percent of the modern figure. That's pretty impressive, and was probably this good only because some of the errors and assumptions made canceled each other out. Still, no flying turtles or giant elephants were needed.

To repeat the experiment, you need to measure the shadow angle of two objects at noon on the same day, and know the distance that separates them. You can work out the angle by measuring the shortest shadow cast, dividing the height of the object by this, and taking the inverse tangent ($\tan^{-1}$) of the answer. Your phone or computer will have a scientific calculator to do this for you, or you can use an online calculator. Alternatively, a spirit level and protractor would do the job.

For Eratosthenes, the distance involved was about 800km (500 miles). A distance of 111km (69 miles) should give you a difference between the two shadow angles of just one degree, which is probably close to the limit of what you might be able to measure. But don't take my word for it – try it!

If that's too much trouble, there are plenty of other ways to easily prove that the Earth is round. Observing ships on the ocean or a Great Lake in calm weather shows that hulls disappear while superstructures and masts remain visible (binoculars will help). The shadow of the Earth on the Moon during a lunar eclipse is always curved. Different constellations are visible from different places. And don't forget that you can see farther from a tall building than you can from the ground.

# Fig. 1 – Suspension bridge curvature

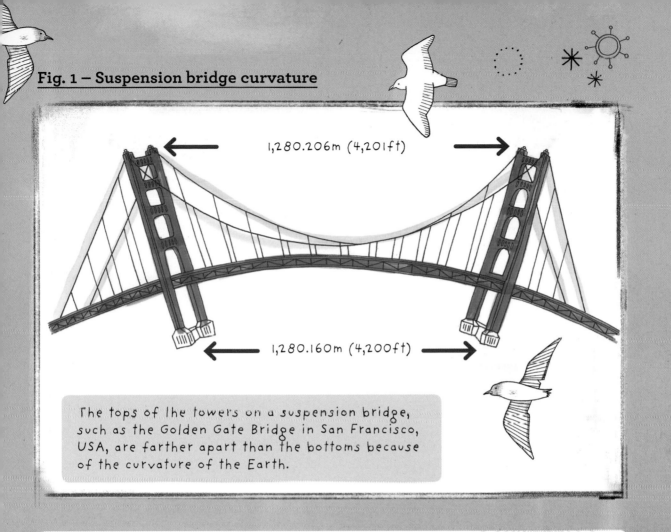

1,280.206m (4,201ft)

1,280.160m (4,200ft)

The tops of the towers on a suspension bridge, such as the Golden Gate Bridge in San Francisco, USA, are farther apart than the bottoms because of the curvature of the Earth.

$$C \text{ (mm)} = \frac{\text{span x tower height (m)}}{6,367}$$

## The C value

My favourite proof of the curved Earth comes from large bridges, although it's not something that's easy to check for yourself. Suspension bridges have two vertical towers, and because of the curvature of the Earth these are farther apart at the top than the bottom. The difference is known as the C value, an informal measure of the monumentality of a structure. It can be calculated by multiplying the span (distance between the towers) by the height of the towers,

both in metres, and dividing the answer by 6,367 (the average radius of the Earth in kilometres). The answer is the number of millimetres the top is farther apart than the bottom.

In England, the Humber Bridge has a C value of 36mm, while the Golden Gate Bridge at San Francisco is 46mm. The current world record holder is in Japan, where the Akashi Kaikyo Bridge has a C value of 93mm.

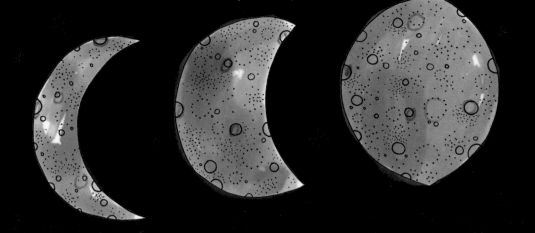

# Lost and found in space

Having conquered the Earth, let's head out a little into nearby space. It is an extraordinary coincidence that the Moon and Sun appear to be almost exactly the same size in the sky, but it is no more than a coincidence (this is also why solar eclipses are so spectacular, but that's a different story).

If we didn't know that the Moon is relatively small and close, while the Sun is large and far away, how could we find out? All you need is a little bit of patience, and an understanding of triangles.

It's fairly obvious to anybody who spends time outdoors that the Moon is essentially spherical, part-lit by the Sun. The bright part of the Moon is always on the sunward side, and the amount of the Moon that is bright changes on a predictable monthly cycle, reaching a maximum when the Sun is opposite in the sky. The basic mechanics of what is going on would be clear even to our prehistoric ancestors.

The waxing and waning cycle means that, twice each month, the Moon is half-lit (slightly oddly, these phases are called first quarter and third quarter). When this happens, the angle of the Sun–Moon–Earth triangle must be 90 degrees at the Moon.

All triangles have internal angles totalling 180 degrees. We now know one of them is 90 degrees, and we are standing on the Earth and so can easily measure the angle here.

It's a really good idea to use a shadow to indicate the line toward the Sun. It would be a really bad idea to try to look anywhere near the Sun itself to measure this angle. Seriously, don't do it.

You can make an angle-measuring device using some card with a short vertical pointer (a gnomon) attached. Draw a line on the card passing through the gnomon. When you see the Moon half-lit (it's almost impossible to tell exactly when this is, to be honest), hold the card up and align it so the gnomon's shadow falls onto the line. Now mark the direction from the gnomon to the Moon. Ta-dah!

With a bit of care, and a bit of luck, you should find that the Sun–Earth–Moon angle is... about 90 degrees (a precision instrument would give about 89.85°). This is an extraordinary result, because it means that the triangle is really long and thin. If the shortest side was one unit, the other two would both be about four hundred. This shows that the Earth and Moon are basically the same distance from the Sun, and that for the Sun to appear the same size as the Moon it must actually be four hundred times bigger.

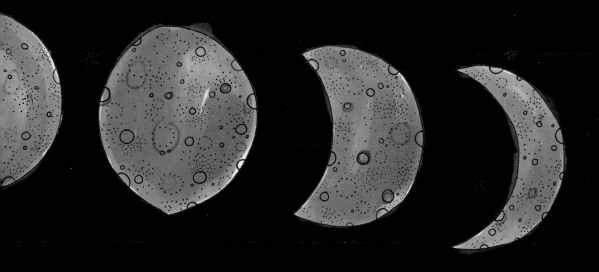

## Fig. 2 – How to measure the distance to the Moon and Sun

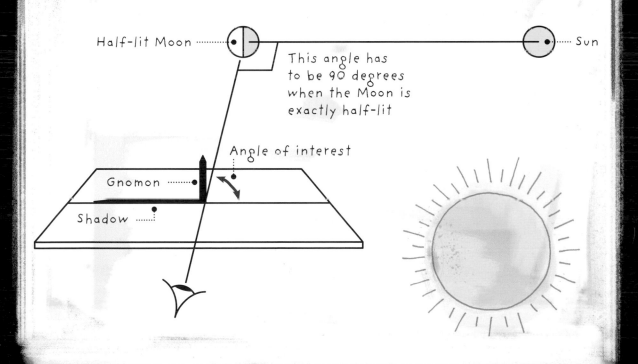

Half-lit Moon

Sun

This angle has to be 90 degrees when the Moon is exactly half-lit

Angle of interest

Gnomon

Shadow

SPACE

# IT REALLY IS ROCKET SCIENCE

Aerospace experts love to say that there's nothing difficult about rocket science, it is the rocket engineering that is the hard part. Certainly, the basics are straightforward, and encapsulated in Newton's third law of motion: for every action, there is an equal and opposite reaction.

In rocketry, that simply means that you arrange for lots of gas to shoot at high speed in one direction, and your rocket will accelerate in the opposite direction.

So, that's easy. We can build rockets and send them anywhere we want, and if we want to go a bit farther we just have to make them bigger, right?

Sadly, it doesn't quite work like that, but there are a number of new and exciting propulsion methods that make use of some improbable-sounding physics, and which might transform the way we power spacecraft.

The problem with conventional chemical rockets is that they have to take their fuel with them, and fuel is heavy. For a rocket to go farther it has to carry more fuel, which is just dead weight until the moment it is needed. It is, in fact, less than useless – it provides no help, but consumes fuel just to be carried.

## Heavy fuel

Airliners on long-haul flights have a similar problem. Ten percent of every tonne of fuel in the tanks is used every hour just to keep itself in the air.

Take the Space Shuttle, with its unusual configuration of a reusable space plane (orbiter) mounted on a huge fuel tank with two external solid fuel boosters. The orbiter's three main engines provided a total thrust at launch of 5,580kN, which was plenty to boost its 760kN weight (77.5 ton mass) into orbit. But it couldn't lift its own fuel! The loaded external fuel tank weighed a further 7,450kN, which is why the solid rocket boosters (12,000kN thrust each) were needed. You could say that the Space Shuttle orbiter blasted itself into space, but only because the **solid rocket boosters (SRBs)** carried its fuel for it.

# Rockets of the future

Escaping the tyranny of chemical rockets is a major goal of future spaceflight, and many possibilities exist. The sci-fi sounding **ion drive** is already a reality and has been successfully used on spacecraft exploring the solar system.

In an ion drive, the fuel and propellant are different things – unlike a rocket, where the energy source (fuel) becomes the reaction mass (propellant).

Ion drives apply high-energy electrons to a stream of gas, ionizing the atoms by knocking electrons off to become a positively charged plasma. Electrically charged grids then attract the plasma before accelerating it away at high speed. The equal and opposite reaction to this acceleration propels the craft forward.

An advantage of ion thrusters is that they can be powered by free solar energy. However, they do require a supply of propellant, usually xenon gas, and so cannot operate indefinitely. They produce a steady continuous thrust, but this is surprisingly feeble and amounts to

a few grams – it is often said that an ion thruster exerts a similar force to the gravity acting on a single sheet of paper on Earth.

Even that is enough for interplanetary navigation, as demonstrated by the Dawn mission. Dawn used three xenon ion thrusters to propel itself first to a Mars fly-by, and then onwards to orbit the giant asteroid Vesta from July 2011. A year later it fired up again to intercept the dwarf planet Ceres, the largest object in the main asteroid belt between Mars and Jupiter.

In the ion engine (see below), atoms of xenon gas are first excited by a magnetic field and then ionized by electrons from the cathode. A high voltage applied between two charged grids attracts and then repels these charged atoms, expelling them at high speed to produce thrust. Finally, a beam of neutralizing electrons is fired into the departing cloud of atoms, to prevent them being attracted back to the spacecraft which would reduce the thrust.

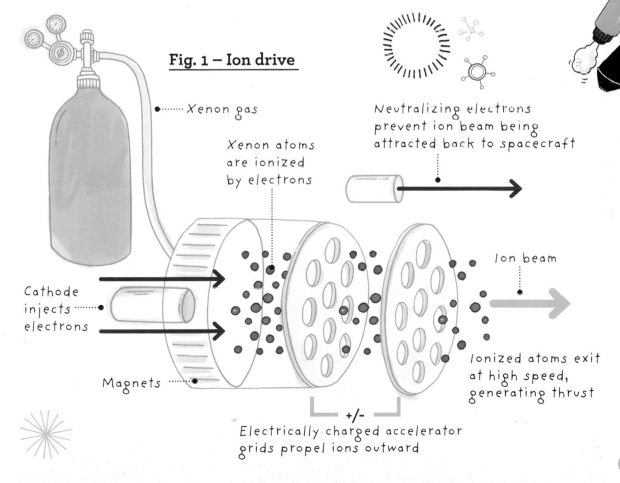

## Fig. 1 – Ion drive

Xenon gas

Xenon atoms are ionized by electrons

Neutralizing electrons prevent ion beam being attracted back to spacecraft

Cathode injects electrons

Ion beam

Magnets

Ionized atoms exit at high speed, generating thrust

+/-
Electrically charged accelerator grids propel ions outward

# We are sailing

Ion drives use free solar power but still require a supply of propellant. What if we could just use solar power alone? Solar sails, which use huge thin sheets of reflective material to generate thrust through the impact of sunlight, may make this feasible.

The physics behind **solar sailing** is not the same as sailing a vessel on Earth, where sails are effectively wings that generate high and low atmospheric pressure, but it does appear quite similar. A spacecraft is boosted by conventional methods and then deploys its sail. This is struck by photons from the Sun, which are mostly reflected backward. This change of momentum generates a small thrust on the sail and any spacecraft attached to it. You could say it is more like flying a kite than sailing a boat.

As with ion engines, the thrust that can be achieved is small, but it is continuous and free. The first, and so far most successful, demonstration of solar sailing began in 2010 with the Japanese Ikaros mission to Venus. This craft unfurled a square sail 14m (46ft) along each side, held taught by being spun at 25rpm and with a small tensioning mass of 0.5kg (1.1lb) at each corner. The push of photons on the sail was experienced as expected, but the craft failed to enter Venus orbit and was given a new mission to sail around the Sun.

There are ambitious plans to use solar sailing to send probes into deep space, including a mission to Alpha Centauri, one of the nearest stars to our own system. The team behind this plan, Breakthrough Starshot, would use powerful space lasers to accelerate a fleet of a thousand probes, boosting them to perhaps 20 percent of light speed at the start of their 20-year mission.

An alternative approach to visiting Alpha Centauri comes courtesy of a rather unlikely sounding propulsion system. The Mach Effect Thruster (MET) makes use of "transient mass fluctuations" which are an effect of special relativity. When a mass is accelerated, relativity tells us that its mass varies by a tiny amount. Working METs have been built on a small scale, using a stack of piezo-electric crystals and two unequal reaction masses to generate small but measurable pulses of thrust.

It is suggested that "impulse engines" using this effect would allow a spacecraft to travel from Earth orbit to the Moon in just four hours, at a constant 1g acceleration. Another proposal would take a probe to Alpha Centauri in 20 years and, unlike the solar sail proposal, this would include time to decelerate and go into orbit rather than simply fly through the alien solar system.

## Fig. 2 – Solar sail

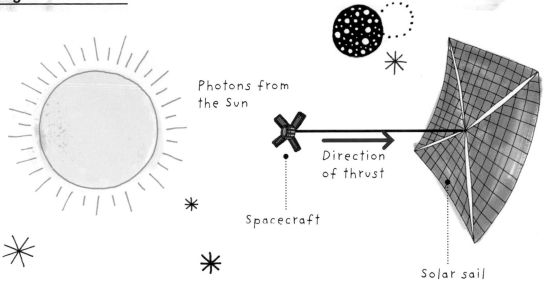

Photons from the Sun

Direction of thrust

Spacecraft

Solar sail

Mach Effect Thrusters seem to contravene one of the fundamental laws of physics, that of conservation of momentum. The explanation is that internal energy changes mean that the fluctuation in mass is due to interaction with the rest of the matter in the Universe, thereby staying within the laws of physics. This is not an easy concept to grasp, but many aspects of relativity and quantum mechanics also seem at odds with everyday reality.

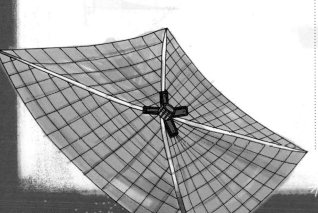

## The improbability drive

One further propulsion method seems the least likely of all, and so has garnered the nickname the **Impossible EM Drive or EmDrive** (more formally a radio frequency resonant cavity thruster). This device, first described in 2001 by Roger Shawyer, consists of a conical cylinder inside which microwaves are bounced back and forth. Conventional theory tells us that this should not produce a thrust, yet laboratory tests show that it does.

Details of a series of NASA experiments published in 2017 showed that the EmDrive consistently produced thrust of 1.2mN/kW. That is only about two percent of the thrust of an ion engine, but it would require no supply of propellant, and is at least two orders of magnitude (100 times) higher than other zero-propellant thrusters such as solar sails.

One of the more plausible explanations for this mysterious effect is known as pilot wave theory, or eurythmic physics, which might also reconcile the bizarre particle/wave duality problem of quantum mechanics. More on that on page 126.

# SPACE WEATHER

We think of weather as the atmospheric conditions we encounter when we step outside, but some of the most dramatic weather takes place in the near vacuum of space. This usually has little impact on our daily lives, but in rare instances the effects can be disastrous.

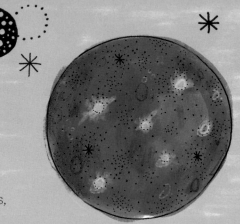

Space weather really means solar weather. The Sun has a cycle of quiet and active phases, lasting about 11 years, with the number and size of **sunspots** peaking at "solar max". Solar cycles have been recorded since 1755 and are numbered sequentially – cycle 24 began in 2008, and delivered the weakest solar max for a century.

The Sun rotates at different rates in different places, fastest at the equator, slowest at the poles, dragging the magnetic field with it. As this becomes wound up, tighter and tighter, the frequency of sunspots increases and their positions move toward the solar equator.

Eventually the magnetic field is drawn so tight that it flips right over, north becoming south and south north. The star relaxes and a new cycle begins.

Sunspots are temporary dark areas of the Sun's surface, a little cooler that the rest of the photosphere (the visible surface) at about 4,000°C (7,000°F) compared to 5,500°C (10,000°F). They are about 1,500–50,000km (900–30,000 miles) across (Earth is 12,742km or 7,918 miles in diameter) and last a few days or weeks. They appear dark only because they shine a little less brightly than the material around them.

Sunspots present no particular risk to us on Earth, but the solar cycle is also associated with two slightly more worrying phenomena – **solar flares** and **coronal mass ejections (CME)**.

## Solar flares

Solar flares are eruptions of broad-spectrum electromagnetic energy. They explode over just a few minutes, releasing radio, microwave, X-ray and gamma radiation, all hurtling into space at the speed of light. It takes about 500 seconds (8 minutes or so) for it to reach us. Earth's magnetic field deflects much of the energy, and the atmosphere absorbs most of the rest, so the main danger is to satellites, spacecraft and astronauts.

Our magnetic field streams energy into the atmosphere near the poles which triggers brighter and more frequent auroras. The northern and southern lights (aurora borealis and aurora australis) are spectacular displays of gas discharge, like a neon advertising sign or fluorescent tube. The high-energy radiation from the Sun strikes molecules of atmospheric gas, causing electrons to perform a quantum jump and release photons.

The wavelength or colour of the light produced indicates what gas has been excited – oxygen glows green and nitrogen glows red and blue (combining to make purple). Because these two gases make up 99 percent of the atmosphere, these are the main colours an aurora produces. The next most abundant gas is argon, at almost one percent, which glows bright violet.

# Coronal mass ejection

More sinister is the coronal mass ejection – a few hundred million tons of plasma from the surface of the Sun, blasted into space. A good analogy is to think of a solar flare as the flash of a gun, and the coronal mass ejection as the bullet.

Thankfully, Earth is a relatively small target for this mass of highly-charged material, and it is unusual for a direct hit to be scored. It takes about three days to travel from the Sun to our vicinity, traveling at about a million miles an hour.

A CME can create a geomagnetic storm resulting in overloaded power and communications networks and satellites fried in orbit. A storm in 1989 caused a huge electrical blackout in Quebec, Canada, while a more powerful event in 1859 produced auroras as far south as the tropics and burned out telegraph networks, delivering shocks to operators and even igniting paper. It's not hard to imagine how much more serious such an event could be today, with our extreme reliance on vulnerable electronics.

## Fig. 1 – Stretched magnetic lines cause solar storms

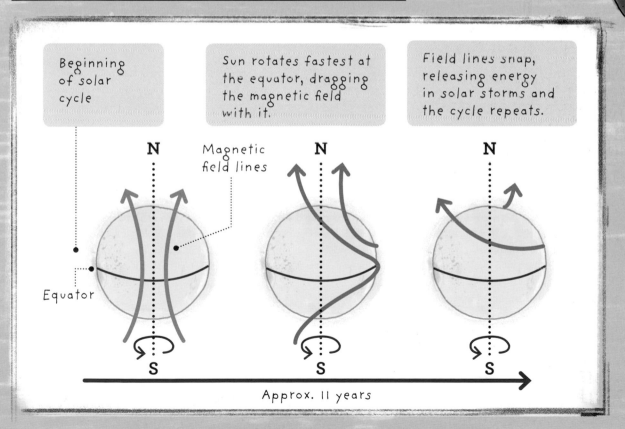

Beginning of solar cycle

Sun rotates fastest at the equator, dragging the magnetic field with it.

Field lines snap, releasing energy in solar storms and the cycle repeats.

Magnetic field lines

Equator

Approx. 11 years

# PLANET NINE

The number of planets in our solar system, by current reckoning, is eight. It hasn't always been eight, not because any have joined or left, but because of changes in what we count as a planet.

The ancient Greeks had five planets, all of which can be seen with the naked eye (Mercury, Venus, Mars, Jupiter and Saturn). In some accounts they included the Moon and Sun as well, making seven.

By the middle ages, with improved understanding of the heavens, we didn't count those two "luminaries" but did include Earth – so that's six planets. The invention of the telescope enabled William Herschel to discover Uranus in 1781, and new understanding of the workings of gravity led to the prediction of Neptune by Urbain Le Verrier, and its discovery by Johann Galle, in 1846. Eight planets.

More mathematics and observations led to the prediction and discovery of Pluto by Clyde Tombaugh in 1930. Planet Nine had been found.

Both Neptune and Pluto were discovered because Uranus kept failing to turn up at its predicted position. We know that orbits are elliptical, not circular, and once a planet's position and speed are determined (this takes as few as three good telescope observations), its course can be predicted. This would work well if there were no other planets to complicate things – but of course there are. What is known as the two-body problem can be solved quite easily, but add another and things get difficult. In the **three-body problem** (for example, Earth, Moon and Sun) each gravitational field interferes with the orbits of the other two bodies.

## Finding Uranus

In the case of Uranus, the bodies to consider are the Sun, Saturn and Jupiter (the inner planets being too small and far away to have any significant effect). Astronomers in the 1840s knew that Uranus was being pulled forward in its orbit by Jupiter, and backward by Saturn. But observations showed Uranus was always late. The most likely reason was that an unknown planet was pulling it backward, and some clever calculations and educated guesswork produced a probable size and position. It was a stunningly good prediction, and Neptune was found that very night.

A similar situation led to the discovery of Pluto, but it was later realized that this world is just too small to have any major influence on the orbits of the giant icy planets. Pluto is tiny, less than a fifth of the mass of our Moon. It was counted as a planet between 1930 and 2006, but is now officially only a dwarf planet. Our tally was once nine, but became eight again. The apparent anomaly in Uranus's orbit was simply an error, and Pluto's discovery merely a coincidence.

# Orbital anomalies

Yet there is still a problem, and some objects seem to miss their appointments in the same way Uranus did. The orbits of six distant asteroids in the Kuiper Belt (where Pluto resides) are strangely skewed, while a tiny discrepancy in the orbit of the Cassini space probe which orbited Saturn from 2004 to 2017 provides another clue. Many astronomers reckon there must be one further large planet way out in the solar system, and have been trying to narrow down its likely position. Their calculations suggest something quite unlike any other object in the solar system.

All of the known planets orbit the Sun in the same plane, more or less, but Planet Nine's orbit is thought to be tilted by about 30 degrees. It's also really far away, in a highly elliptical orbit lasting around 15,000 years. It may come as close as 200AU and travel as far as 1,200AU from the Sun (at least ten times farther out than Pluto). It is thought to be about the size of Neptune, and ten times the mass of the Earth, but nobody has seen it yet, or can be completely sure it exists. Planet Nine is sometimes called Planet X but, if it really is the ninth, that should be Planet IX.

An AU (astronomical unit) is the average distance between the Earth and the Sun, about 150 million kilometres or 93 million miles.

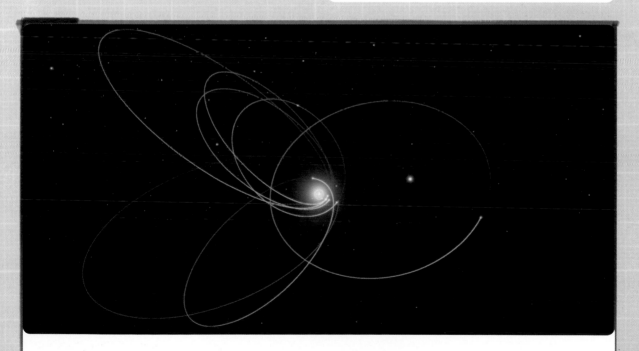

The strange lop-sided orbits of six distant asteroids in the Kuiper Belt, shown in magenta, could be explained by Planet Nine.

# SPACE TELESCOPES

Astronomers face many obstacles, but from the surface of Earth one dominates their observations: the atmosphere.

It might appear nice and transparent, on a good day at least, but it actually absorbs a good deal of the visible and other wavelengths that enter it. Pollution, clouds and scattered light from cities add to the problem, and so the best observatories are located on mountaintops in dry remote areas.

Going up high can put most of the problem below you. The world's highest observatory, the University of Tokyo Atacama Observatory in Chile, is at 5,640m (18,500ft) which is above more than half of the atmosphere. But it means that anybody working there, or building it, needs a continuous oxygen supply or they might be seeing stars because of altitude sickness alone.

Pilots of unpressurized aircraft often use oxygen above 10,000ft, and have to above 12,500ft (3,810m).

## 'Scopes in space

What if a telescope could be above *all* of the atmosphere?

The Hubble space telescope is the most famous observatory in space, and (after a few embarrassing teething troubles that required in-orbit repair) has returned many spectacular images and helped make many important discoveries. It has been doing this since the 1990s and should continue at least into the 2020s.

Hubble's instruments work in visible light, near-ultraviolet and near-infrared wavelengths. The James Webb space telescope (JWST), the next great astronomical instrument, is a different beast altogether. Not only is it much bigger, with more than seven times the mirror area, but it will work primarily in infrared. This will allow it to detect the tiny residual energy from the beginning of the Universe, which is at wavelengths that cannot be detected from the surface of Earth.

It also means that, despite being positioned in the freezing void of space, some of JWST's instruments actually have to be cooled in order to work.

The whole telescope assembly is shaded by five layers of parasol, each about the size of a tennis court, designed to reflect radiant heat energy from the Sun. Space itself may be at a temperature of only about 3 kelvin (3K, three degrees above **absolute zero** or about -270°C, -455°F), but without the sunshield the telescope would begin to heat up by absorbing radiation.

# Imaging in infrared

Because JWST is collecting and analysing infrared wavelengths, it could be considered a very fancy thermometer. Infrared is basically heat energy, and the telescope will measure small differences in the temperatures of very cold objects far away in the distant Universe or in places like the Oort cloud on the edge of our solar system. It can only do that if it is itself cooler than the objects being studied.

While the hot side of the telescope will reach a scorching 313K (85°C, 185°F), the sunshield's passive cooling keeps the cold side of the telescope below 40K (-233°C, -388°F). This is enough for the near-infrared detectors, but a mid-infrared instrument called Miri needs to be as cold as 7K (-266°C, -447°F) before it can work. To chill it, two refrigeration units are needed. A pulse tube precooler pumps heat away by compressing and decompressing helium gas, bringing the detector down to 18K, and then a Joule-Thomson loop heat exchanger takes it down to 7K by allowing helium to expand and therefore cool.

## Kelvin scale

The lowest possible temperature in the Universe is absolute zero, at which point all atomic motion and vibration ceases. This temperature is also the starting point for the kelvin scale, which is used because it is more convenient than having lots of negative numbers to deal with (there is no such thing as a minus kelvin temperature). One kelvin equals one degree Celsius, and 0°C is 273.15K.

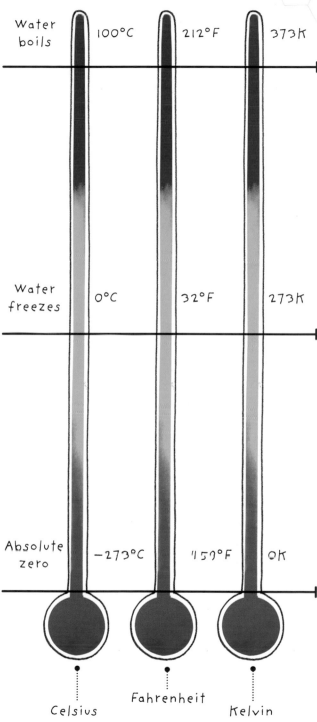

# Fig. 1 – Transparency of Earth's atmosphere at different wavelengths

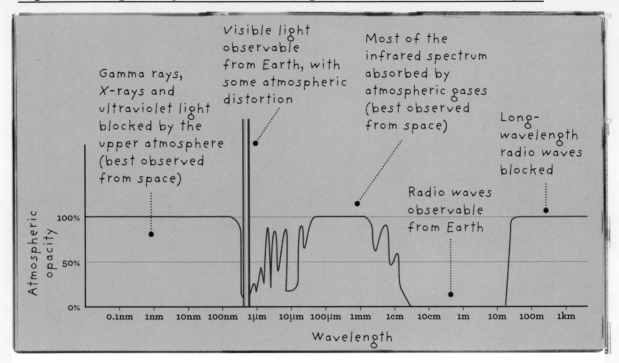

Gamma rays, X-rays and ultraviolet light blocked by the upper atmosphere (best observed from space)

Visible light observable from Earth, with some atmospheric distortion

Most of the infrared spectrum absorbed by atmospheric gases (best observed from space)

Long-wavelength radio waves blocked

Radio waves observable from Earth

Atmospheric opacity

100%

50%

0%

0.1nm  1nm  10nm  100nm  1μm  10μm  100μm  1mm  1cm  10cm  1m  10m  100m  1km

Wavelength

## Transparent evolution

Earth's atmosphere is quite transparent to some wavelengths of the electromagnetic spectrum, and completely opaque to others. It might seem strangely convenient that visible wavelengths – the light we see by – can pass through, but this is no coincidence. We evolved to see with the available light.

Within the visible range, the wavelength at which the atmosphere is most transparent is about 475nm, or "sky blue". There are further windows of partial transparency in some near-infrared (NIR) bands. The main transparent window of the atmosphere is actually in the radio spectrum. This means radio telescopes can be built at convenient lower altitudes – the famous Jodrell Bank radio telescope in Cheshire, UK, is just 77m (250ft) above sea level.

nm = nanometre = billionth of a metre

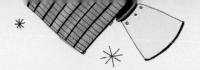

# Ten gold rings

None of this would be of any use without the primary mirror, which collects and focuses the faint infrared energy of the deep space objects being studied. The most obvious thing about JWST is that this mirror is made of 18 hexagonal gold segments, giving it a honeycomb appearance.

The hexagonal arrangement was chosen for engineering reasons, as the mirror has to be packed up to fit onto a rocket for launch. The gold, though, was chosen because of its special reflective properties.

A perfect reflector should appear colourless, because any tint means that some wavelengths have been absorbed instead of reflected. But JWST is only interested in collecting infrared wavelengths, and gold offers excellent reflective qualities for these.

Besides being expensive, gold is too heavy and soft to make such a large structure. The hexagonal panels are actually made of beryllium – a light metal similar to aluminium but with better thermal stability. The gold is merely a fantastically thin coating, just a ten-thousandth of a millimetre thick, applied to the beryllium.

This tiny amount of gold is about the same as ten wedding rings, yet is the critical part of the whole telescope. It is what does the actual work of collecting infrared light. Everything else – the sunshield, the hexagonal panels, the rocket launch – is dedicated to placing these 48g (1.7oz) of precious metal in space.

## Fig. 2 – James Webb Space Telescope

The James Webb Space Telescope's main mirror has a thin coating of gold and systems to cool its sensitive instruments below even the extreme cold of outer space.

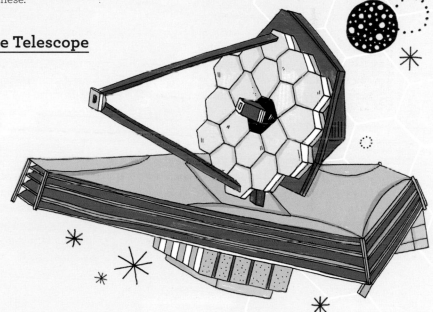

# Keeping cool in space

Spacecraft including the International Space Station (ISS) and the Apollo missions have a bigger problem staying cool than staying warm. The vacuum of space means heat cannot be dumped by convection or conduction, but only by radiation.

Your home or office may be heated by radiators, but really these should be called convectors because they transfer most heat by convection. Hold your hand close to the bottom, then close to the top, and notice the difference.

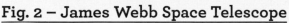

# IS THERE ANYBODY OUT THERE?

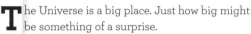

Perhaps the greatest unanswered question in human existence is whether we are alone in the Universe. Are there other intelligences out there somewhere, perhaps also scouring the heavens in search of somebody to have a conversation with? Or are we the only creatures, anywhere, capable of even asking that question?

The Universe is a big place. Just how big might be something of a surprise.

We know it all started with the Big Bang, around 14 billion years ago. We know that nothing can travel faster than the speed of light. It would seem logical that the Universe cannot be more than around 28 billion light years across.

If we see an object near the limit of the observable Universe, 14 billion light years away, we see it as it was 14 billion years ago.

Modern telescopes allow us to see almost, but not quite, to the edge. Distance makes objects appear dim, and the speed of divergence caused by universal expansion causes light to appear shifted to the red end of the spectrum. This red shift can ultimately give the photons a wavelength so long they become difficult to detect.

Taking the expansion into account, best estimate for the actual size of the Universe today is around 46 billion light years across. It will be a little more tomorrow.

Inside that unimaginable volume might be as many as ten trillion (10,000,000,000,000) galaxies, each made of perhaps 100 billion stars. That would make a universal star count of a trillion trillion, or a septillion; 1 followed by 24 zeroes, and at least 100 times more than the number of grains of sand on all the beaches, deserts and oceans of planet Earth.

If we assume that most stars are orbited by planets and moons, the number of parcels of celestial real estate must be what scientists like to call A Large Number.

Is it conceivable that in that vastness, on those uncountable alien worlds, life only exists on the one insignificant blue-green orb we call home?

## The Drake equation

Our galaxy, the Milky Way is a spiral disc about 100,000 light years across. What are the chances that life exists somewhere else in this, our own neighbourhood? And what are the chances that there is *intelligent life*, able to exchange signals and maybe even visits?

This is the basis of the famous **Drake equation**, created by astronomer and astrophysicist Frank Drake in 1961. He wanted to work out how many extraterrestrial civilizations might exist, and adopted a systematic approach to do so. He thought there were seven factors to consider, starting with the rate of star formation and ending with the length of time an intelligent civilization exists:

$$N = R^* \times f_p \times n_e \times f_l \times f_i \times f_c \times L$$

**N** is the number of detectable civilizations. The other factors are the rate of star formation ($R^*$), the fraction of stars that have planets ($f_p$), the average number of planets in the habitable zone ($n_e$), the fraction of those that develop life ($f_l$), the fraction of those that produce intelligent civilizations ($f_i$), the fraction of those that develop communications technology ($f_c$) and finally the number of years the civilization survives ($L$).

So what's the answer? The biggest problem with the equation is that almost every factor involves making a wild guess. When Drake first ran the numbers, answers ranged from a thousand to a hundred million.

A lot has changed in our understanding of the galaxy in the decades since. Then, nobody was sure whether planets even existed beyond our solar system, but we now know that they are common. The first exoplanet (a planet outside our solar system) was discovered in 1992, and with every improvement in technology we find more. Today there are thousands of known worlds orbiting distant suns.

## Habitable zones

Another big change is that we no longer think only of planets as possible homes for life, but moons as well. In our system there are three planets that might be in the habitable zone. Venus is today too hot, and Mars probably too cold, but with different atmospheres either could support life. There are also a handful of moons orbiting Jupiter and Saturn that have liquid water, and Titan (Saturn's largest moon) has a thick atmosphere that is rich in organic compounds.

All of this looks promising for those hoping to establish first contact. But one big question remains: where is everybody? If there are so many great places for life to evolve, why don't we have any solid evidence?

The question is known as the Fermi paradox, after physicist Enrico Fermi (most famous for building the first nuclear reactor) who pointed out the contradiction.

Life on Earth seems to have begun almost as soon as the planet cooled to a reasonable temperature. The human species *Homo sapiens* has existed for around 200,000 years, and our intelligent civilization, such as it is, began around 3,200 years ago with agriculture and cities. Communications technology began with the invention of radio by Guglielmo Marconi in 1899.

This means that the very first human radio signals to leak into space, feeble and undirected, have been travelling for around 120 years. For an advanced civilization to have detected and responded to this, and for us to have heard their response, ET would have to be no more than 50 light years away. There are just 133 known stars within this bubble, out of 100 billion in the galaxy.

In other words, it would be a one in a billion chance if this had happened.

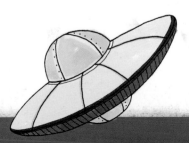

# Wobble and dip to find ET

The two main techniques used to find exoplanets involve looking at changes in distant stars, and they work best at detecting large planets in close orbits.

A planet's gravity pulls on its star, so that both orbit around their common centre of mass or barycentre (see Life on Earth, page 90). This causes a wobble that can be detected as a small change in the speed of the star moving toward or away from Earth. The light from the star will show a small Doppler shift, changing on a regular cycle indicating the orbital period of the planet. Viewed from afar, the Sun would show this change in radial velocity mainly because of the giant planet Jupiter, with an orbit taking nearly 12 years.

The barycentre of the Sun and Jupiter is actually outside of the Sun's visible surface (the photosphere).

In cases where a planet's orbit takes it across the face of its star as seen from Earth, it may cause a measurable dip in brightness. It is, essentially, an eclipse of a distant star. The dip in brightness shows quite sharp edges on a graph, and the amount of dimming provides an estimate of the diameter of the eclipsing planet. Some of the light that reaches us will have passed through the atmosphere of the planet, if it has one, and this raises the exciting possibility of being able to detect oxygen and organic molecules. This might provide the first indirect evidence of alien life.

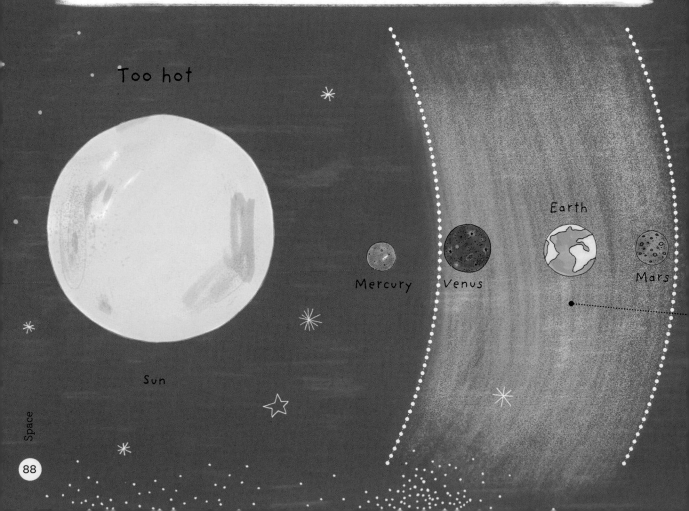

Too hot

Sun

Mercury

Venus

Earth

Mars

# Listening for life

There have been a number of projects to listen for radio signals that might originate from alien intelligences, known collectively as SETI – Search for Extraterrestrial Intelligence. Their lack of success so far might have something to do with how we are listening.

Many early attempts concentrated on the radio frequency of 1,420MHz (21cm wavelength), which is the natural frequency of hydrogen, the most abundant element in the Universe. It was thought that this specific band might be a natural choice for aliens, because it was so universal. It is just as likely that they might avoid this waveband because a signal could be lost in the naturally occurring radio noise of the cosmos.

Radio communications might not be much used by advanced technologies. They may use lasers, or entangled particles, or some other technology we can't even conceive of. Perhaps they look on radio in the way we view smoke signals, as a quaint and rudimentary form of communication. And perhaps they are completely aware of us, but have chosen to remain incognito.

The Breakthrough Listen project is the latest systematic attempt to find ET via radio and laser. It involves surveying a million of the nearest stars to us, across a wider part of the radio spectrum, and with greater sensitivity than ever before. Stay tuned!

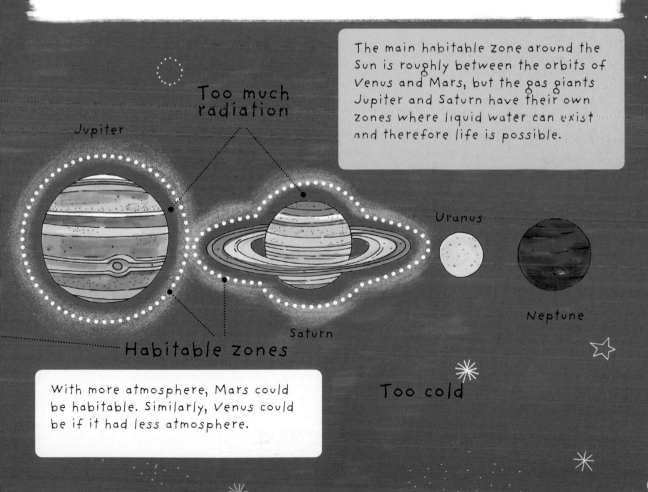

The main habitable zone around the Sun is roughly between the orbits of Venus and Mars, but the gas giants Jupiter and Saturn have their own zones where liquid water can exist and therefore life is possible.

Too much radiation

Jupiter

Uranus

Neptune

Saturn

Habitable zones

With more atmosphere, Mars could be habitable. Similarly, Venus could be if it had less atmosphere.

Too cold

# LIFE ON EARTH

The pale blue dot we call home is, so far, the only place in the whole Universe where we know life exists. A specific set of circumstances here allowed organic molecules to begin joining together and making copies of themselves, slowly changing and evolving into everything from a humble bacterium to something as complicated and awesome as you and me.

Earth spins on axis in one day

The question is, which of the peculiar features of Earth were important in this? All life on Earth needs liquid water to survive, so our orbit around the Sun at a distance that is not too close, not too far away, must play a part. We are in the so-called **Goldilocks zone**, just the right distance from the Sun so the surface is not too hot or too cold. However if the atmosphere was thinner, we would need to be closer to the Sun, and if it was thicker, we'd have to be farther away, so conditions for life do not just depend on the distance from the Sun.

## Shields up!

Our planet's strong magnetic field probably plays a part too. It's like a force field which deflects dangerous radiation and high energy particles from the Sun. Of the four rocky inner planets in our system, only Earth has this. Mercury has a little (about one percent of Earth), but both Venus and Mars have practically zero magnetic field. Jupiter has a field about 20,000 times as strong as ours.

A planet with the right atmosphere at the right distance, but no magnetic field, *could* develop life. But it might be that the bombardment of radiation would be too big an obstacle. We don't know.

The Moon could also be important. Having such a large companion means the Earth's oceans are pulled and dragged around the surface of the planet, with tides creating inter-tidal zones that are sometimes wet, sometimes dry. It might be that this was where biology first took hold.

Moon's centre of mass also orbits the barycentre

Moon

Centre of mass of Moon

Earth

Centre of mass orbits the barycentre

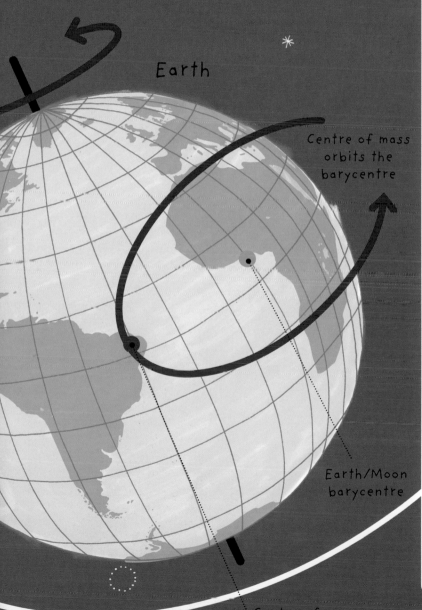

Earth/Moon barycentre

Centre of mass of Earth

## Is Earth a double planet?

Earth has the largest moon in the solar system, by proportion. No other satellite has a mass more than about 1/4,000th of its planet, but ours is 1/81st. It's easily the largest on this comparison, although some others are bigger in absolute terms.

Our Moon is so big that it doesn't really orbit the Earth, but rather both orbit the common centre of mass of the Earth/Moon system. This is the barycentre, and it is closer to the surface of the Earth than it is to the centre. If the Moon was directly overhead, the **barycentre** would be about 1,700km (1,000 miles) beneath your feet, moving at 40km/h (25mph) approximately to the east keeping pace with the changing position of the Moon. Meanwhile the ground you are standing on is moving at 1,000km/h (600mph) – depending on your latitude – due to the daily rotation of the planet.

The Earth/Moon barycentre is inside the Earth and, to be considered a true double planet, some astronomers think the barycentre should be in open space. Pluto and its companion Charon meet this requirement, but as Pluto is no longer officially a planet it cannot be a double planet.

# LIFE ON MARS

The fourth rock from the Sun does not have a thick atmosphere, a strong magnetic field or a large moon, all of which may have been important to life getting started on Earth (see **Life on Earth**, page 90). But we know from the hugely successful Mars rovers that the planet did once have the most important feature for life – liquid water.

Whether this was enough for biology to get going, and if so whether any of it is still alive, are two of the biggest questions in space exploration today. One day a robot, or perhaps a human explorer, may crack open a rock and find a fossil. That would answer the first part. The second probably depends on whether liquid water can still exist on the red planet.

Mars is just about close enough to the Sun to get enough warmth, but being within the Goldilocks zone isn't enough. The planet is so cold, and its atmosphere is so pathetically thin, that liquid water should not be able to exist at the surface. Probably.

Spacecraft orbiting Mars have returned tantalizing images showing patches which appear to get darker with the Martian summer, almost as if they were wet. The intriguing possibility is that water may exist beneath the surface, locked up in ice which melts periodically, or may condense from the atmosphere to form rivulets.

The temperature and pressure conditions needed for water to be either solid (ice), liquid or gas are revealed by a **phase diagram**. Notice in particular the **triple point,** a special set of circumstances where ice, liquid or vapour can exist.

Intriguingly, the pressure at the surface of Mars is about the same as the triple point. This means that there may be a small range of temperatures close to 0°C (32°F) where liquid water could exist. If the surface pressure could be raised, perhaps by some future geoengineering, that range would increase significantly.

The possible existence of liquid water at the surface does not mean that life exists. These are very tough conditions, with an average temperature of -55°C (-67°F). But it could be one less reason why not.

It might be significant that the phase diagram is for pure water, rather than salty. The Mars rovers have revealed many salts in the planet's crust, including calcium and magnesium perchlorate, so any water could well be briny – and liquid over a wider range.

Dark streaks in the Martian soil reappear each summer, and indicate flowing liquid, which might be salty water. Image from Mars Reconnaissance Orbiter.

## Fig. 1 – Phase diagram for water

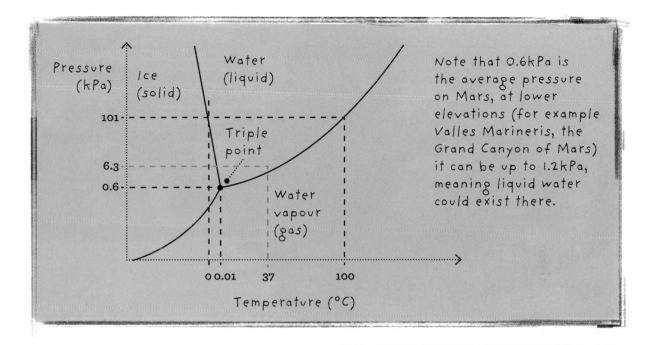

Note that 0.6kPa is the average pressure on Mars, at lower elevations (for example Valles Marineris, the Grand Canyon of Mars) it can be up to 1.2kPa, meaning liquid water could exist there.

## The Armstrong limit

The Armstrong limit is an atmospheric pressure so low that water boils at or below human body temperature (37°C or 99°F). This means that the fluid in a person's eyes, mouth and lungs would evaporate, and their blood would literally boil away. On Earth, this occurs at an altitude of around 18,000m (60,000ft), corresponding to a pressure of 6.3kPa (the pressure is about 100kPa at ground level).

On the surface of Mars the pressure is only about 0.6kPa – not even one tenth of the Armstrong limit. Forget what you may have seen in the movies, the laws of physics mean there is no way for a human to survive on Mars without an operational spacesuit.

## Do Martian bugs fart?

Curiosity Rover detected a rapid increase in the methane content of the Martian atmosphere in late 2013 and early 2014, causing much excitement among scientists. Methane is an organic molecule, which only means it contains carbon, not that it comes from an organism. It also breaks down pretty quickly, so an increase means the gas is being generated. But how?

About 90 percent of methane on Earth is from life (flatulent cows are one significant source). The gas detected on Mars could be from micro-organisms waking up in the short summer, but it could also be from geological sources.

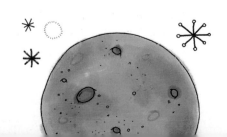

BIG SCIENCE

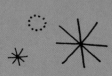

# WHAT'S THE MATTER?

Although matter and energy seem like very different things, one of the core revelations of Albert Einstein's theories of relativity is that they are really just alternative versions of each other. Matter is basically concentrated energy, and one can be converted into the other.

That's not to say that they are *easily* interchangeable, but it can be done. The nuclear fusion reactions that power the Sun convert 620 million tonnes of hydrogen into about 616 million tonnes of helium every second of every day, releasing unimaginable amounts of energy in the process. By the time it reaches the daytime surface of the Earth, the power is about the same as a one-bar electric fire (1kW) on every square metre.

The Sun has been doing this for about 4.6 billion years, and is expected to continue for about another 5 billion before ballooning into a red giant and obliterating Earth. As the hydrogen becomes converted into helium, the Sun will start to use this as fuel. Helium is denser than hydrogen, so the nuclear reactions occupy a smaller volume. This makes the outer layers of the Sun hotter, so they will expand and destroy the Earth.

There are a number of attempts to harness fusion power here on Earth, which would provide an almost limitless supply of cheap non-fossil energy without problems of radioactive waste disposal. The technology remains decades away from commercial availability.

## Einstein's equation

The world's most famous equation, $E = mc^2$, reveals how much energy you can get from a given amount of matter. Energy equals mass times the speed of light squared. We know that the speed of light is a big number, whatever system of units you use, and a big number squared becomes a very big number indeed. This explains how a tiny amount of matter converts into a huge amount of energy.

## Catch a falling Tsar

The most dramatic demonstration of mass-energy equivalence on Earth is the detonation of a nuclear weapon. The uranium bomb dropped on Hiroshima in 1945 released the energy equivalent of 15,000 tonnes of TNT, by converting just 0.7g (0.02oz) of matter into energy. That's about a sixth of a teaspoon of sugar.

The so-called Tsar bomb (RDS-220 hydrogen bomb), built by the Soviet Union in 1961, is the most powerful weapon ever tested. It weighed 27 tonnes, and when detonated converted just 0.01 percent of its mass – 2.67kg (5.9lb) – into the energy equivalent of 50 million tonnes of TNT. All the conventional explosives used in the Second World War are reckoned to add up to only 5 million tonnes of TNT.

## Fig. 1 – The fusion furnace of the Sun converts hydrogen to helium plus energy

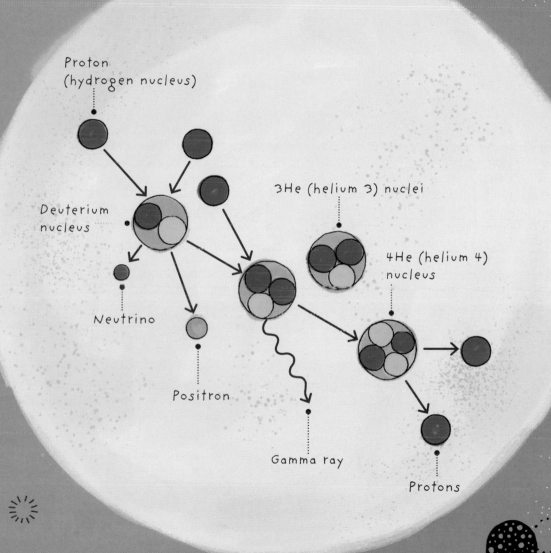

Proton (hydrogen nucleus)

Deuterium nucleus

3He (helium 3) nuclei

4He (helium 4) nucleus

Neutrino

Positron

Gamma ray

Protons

# TIME TRAVEL

Can we travel in time? The flippant answer is an obvious "yes". We're all traveling forward in time at the rate of one second per second. More interesting is whether we can travel faster or slower into the future, or backward to the past.

**R**elativity tells us that time and space are intimately intertwined into the four-dimensional fabric of spacetime. Events occur at a physical location and time, and two events cannot exchange information any faster than the speed of light. We can travel freely in the three physical dimensions, so what about the fourth? Any movement in space also involves movement in time, because it takes time to get from one place to another. That means that time is changing for any moving object.

If that moving object was a clock, it would appear to run slow from the perspective of a stationary observer. This is the critical observation in Einstein's special theory of relativity. This **time dilation** means that the time experienced by two observers can be stretched if they are moving relative to each other at high speed (it actually applies at any speed, but the effect is tiny until traveling at a significant proportion of light speed). To get time to run at half speed, you would need to reach about 86 percent of light speed. One-third time speed would require about 94 percent light speed, and one quarter about 96 percent.

There are lots of examples which confirm this perplexing situation. A favourite is the "twin paradox" where one identical twin is blasted into space while a sibling remains on Earth. When they reunite, the traveling twin should be a little bit younger. Amazingly, this has actually happened. Scott and Mark Kelly are identical twin brothers from New Jersey, who both became NASA astronauts. Scott spent 520 days in orbit, but Mark only 54. This means that Scott has aged five milliseconds less than his big brother.

## Fig. 1 – Time dilation

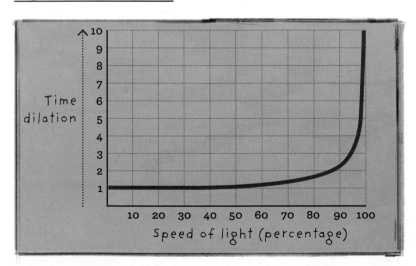

Time dilation

Speed of light (percentage)

Our normal experience is undilated time (dilation = 1), but as you travel faster it increases and would be infinite at the speed of light

$$t' = t \sqrt{1 - V^2 / c^2}$$

Where: t' = dilated time
t = stationary time
V = velocity
c = speed of light

## Fig. 2 – Grandfather paradox

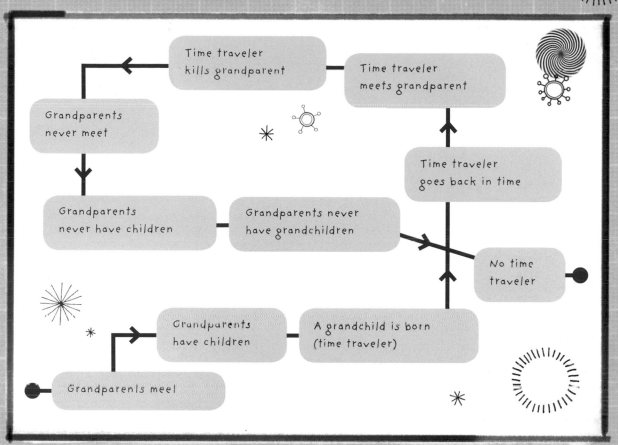

Time traveler kills grandparent

Time traveler meets grandparent

Grandparents never meet

Time traveler goes back in time

Grandparents never have children

Grandparents never have grandchildren

No time traveler

Grandparents have children

A grandchild is born (time traveler)

Grandparents meet

# The grandfather paradox

The real question is whether the laws of physics as we understand them would allow or forbid traveling backwards in time. Notice that I said "as we understand them", because this is one of many areas in advanced theoretical physics for which the only sensible answer is "maybe, but we don't really know".

One reason why some people think it must not be possible is known as the grandfather paradox. Say you invent a time machine, travel back to the day before your grandparents meet, and do something horrible such as killing one of them. They never meet, never have children (one of whom is your parent) and so you can never be born. Therefore you cannot have built the time machine to cause all this chaos.

Does this feat of logic mean time travel is impossible? Science fiction writers regularly conjure explanations to sidestep the issue, but what about physicists?

General relativity explains that gravity is the result of the warping of spacetime by matter and energy. It implies that if you had enough of those, you could generate enough gravity to fold spacetime back on itself – a situation called a **closed timelike curve** or CTC. That would, in theory, allow a traveler to loop backward in time.

So where are these time travelers? One argument that it can't be done is that we have no evidence that it has been done, but this is like trying to prove a negative. Relativity does not expressly forbid backward time travel, so it seems that it is not impossible, just really, really difficult.

# BOMB OR MELTDOWN?

It's a situation I hope you never face, but if you had to choose between being in a nuclear blast or a nuclear power station failure, which would be better? Option three, I hear you shout: far far away from either of them. Fair enough.

At the time of writing, nuclear weapons have been used in anger twice – by the USA against the Japanese cites of Hiroshima and Nagasaki in August 1945. The Hiroshima uranium bomb killed around 60,000 people immediately and at least 100,000 within months. The Nagasaki plutonium bomb was more explosive – 21 kilotons compared to 16 – yet killed fewer due to the hilly terrain and perhaps being a couple of miles off-target. It was still horribly destructive with 40,000 killed outright and 80,000 by the end of 1945.

Major nuclear power station incidents have occurred more. There have been 33 serious nuclear power accidents since 1952, which the International Atomic Energy Authority (IAEA) ranks on the International Nuclear and Radiological Event Scale (INES) from 1 (anomaly) to 7 (major accident). By far the most serious, and the only ones to be ranked at level 7, are the Chernobyl disaster in Ukraine (then part of the USSR) in 1986 and the Fukushima Daiichi disaster in Japan in 2011. Within a few days of Chernobyl 31 people had died and it is estimated that the disaster will cause 4,000 deaths through cancer and 40,000 extra cases of cancer. Although thousands of people were killed in the evacuation and as a result of the earthquake and tsunami that triggered the incident, Fukushima caused no immediate deaths through radiation. Here, estimates of future deaths by cancer range from zero to a hundred or so.

So far, your odds look a lot better at a failing power station than a bomb site. But it isn't that simple.

Nuclear weapons kill people quickly by blast, and then slowly by radiation poisoning. A bomb detonated on the ground leaves a huge crater, and all the matter that used to be there is vaporized to become radioactive fallout (tiny, solid, deadly particles). Both the Hiroshima and Nagasaki weapons were air-burst – detonated above ground – which means not much material was vaporized to become radioactive fallout. Chernobyl, by contrast, was detonated at ground level which meant that hundreds of tonnes of material was made highly radioactive and sent high into the sky. It rained down across a huge stretch of Europe from Russia to Ireland. Some estimate that this disaster released between 100 and 400 times as much radioactive material to the atmosphere as the Hiroshima and Nagasaki bombs combined.

## Radioactive half-life

The main harmful fallout at Hiroshima was uranium-235, with a half-life of 700 million years. At Nagasaki it was plutonium-239, with a half-life of 24,000 years. To all intents and purposes, those elements are just as radioactive today as they were in 1945.

At Chernobyl and Fukushima, the primary concern is caesium-137. This has a much shorter half-life, only 30 years, meaning that at Chernobyl the caesium contamination has already reduced by more than half.

Even so, the sheer level of contamination means that anywhere within about 30km (20 miles) of the power station will be uninhabitable for centuries. Even the official clean-up operation is not scheduled to be completed until 2065.

In contrast, both Hiroshima and Nagasaki are today thriving cities with no higher background radiation than before the bombs.

## Fig. 1 – Half life

| Incident | Radionuclide | Half-life |
| --- | --- | --- |
| **Hiroshima** | uranium-235 | 700 million years |
| **Nagasaki** | plutonium-239 | 24,000 years |
| **Chernobyl** | caesium-137 | 30 years |
| **Fukushima** | caesium-137 | 30 years |

## What is a meltdown?

Nuclear reactors exploit the fission reaction in which atoms of a heavy radioactive element split into two, releasing energy. A coolant such as water becomes heated, turning into steam to power electricity-generating turbines.

Careful control of reactor temperatures is vital. If coolant levels fall, or other control methods such as reaction-quenching rods fail, the nuclear fuel can become hot enough to melt itself. It will then pool in the bottom of the reactor, continuing to generate heat and perhaps melting the containment vessel. This is a meltdown, or officially a "core melt accident".

Whatever you call it, it is a Very Bad Thing.

## Fig. 2 – Nuclear reactor

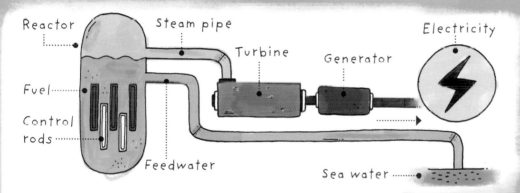

Reactor · Steam pipe · Turbine · Generator · Electricity · Fuel · Control rods · Feedwater · Sea water

## Fig. 3 – Meltdown process

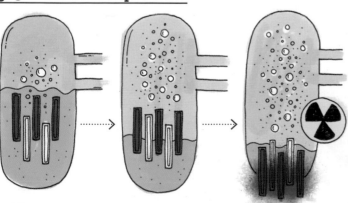

If the cooling system fails, the nuclear fuel continues to heat up, boiling off any remaining coolant and increasing pressure so that the reactor vessel is in danger of bursting. As fuel temperatures rise further the rods themselves begin to melt, pooling in the bottom of the reactor and melting through the base.

Big Science

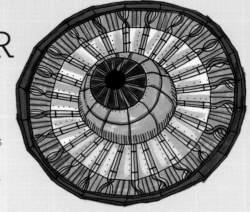

# THE LARGE HADRON COLLIDER

The biggest and most powerful machine in the world, the Large Hadron Collider (LHC) is dedicated to finding and measuring the smallest things in the cosmos. It does this by accelerating atomic nuclei up to very nearly light speed, smashing them into each other, then picking over the remnants of the collisions.

The LHC is actually several particle accelerators joined together, the main part being a circular tunnel 27km (17 miles) in circumference. This sits about 100m (330ft) below ground, under the border between France and Switzerland. This vast instrument is so sensitive that allowance is made for the effect of a full moon, which changes its length by just 1mm.

Hadrons are subatomic particles, the most familiar of which are the protons and neutrons that make up an atomic nucleus. In the LHC, beams of nuclei of hydrogen, lead or xenon are smashed into each other to glimpse what they are made of and recreate conditions similar to the very early Universe.

## Protons in a bottle

Most experiments just use protons, sourced from a standard bottle of hydrogen gas, which lasts about six months. Hydrogen is the atomic building block of the Universe, with a single electron and a single proton. The gas is released into the high vacuum of the LHC, first passing through a series of ionizing magnets which strip away the electrons to leave just the protons.

These are accelerated over several hours by a series of superconducting electromagnets operating at -271°C (-456°F) – colder than space. In the process, the proton beam is divided into 2,808 distinct bunches, each a few centimetres long and containing a hundred billion protons.

Beams are sent around the accelerator ring in opposite directions, each reaching 99.9999991 percent of light speed, and focused to just 20µm across (less than half the width of a human hair). Then the beams are crossed.

Two hundred billion hadrons fly into each other, and for the most part miss altogether. Only about 40 collisions occur, but these tiny flashes of energy reveal tell-tale traces of the shadowy sub-hadron realm. The amount of data generated is staggering, and expressed in a delightfully idiosyncratic unit: in 2017, LHC recorded 5 quadrillion collisions, producing 50 inverse femtobarns of data.

The inverse femtobarn is a measure of both the number of collisions and the amount of data collected. 1fb⁻¹ is about 100 trillion proton–proton collisions.

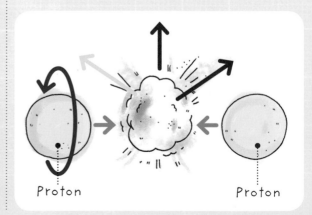

Proton          Proton

The most famous discovery at LHC is confirmation of the existence of the Higgs boson, one of the fundamental particles in the particle zoo of the Standard Model (see The Standard Model, page 106). This boson (force-carrying particle) was predicted in 1964 by Robert Brout, François Englert and Peter Higgs as the mechanism to give mass to other particles. It was finally found at LHC nearly 50 years later, in 2012.

The LHC may one day be superseded by an even more enormous supercollider called the Future Circular Collider. This would have a circular tunnel three or four times as long (80–100km, or 50–60 miles), with correspondingly higher energies, to allow physicists to glimpse even deeper into the subatomic realm.

## High energy collisions

The energy of particles screaming around the LHC is measured in tera-electron volts, with each proton gaining a maximum energy of 7TeV – so that a smack-on collision energy is 14TeV. Lead-ion beams, being much more massive, have collision energies up to 1,150TeV.

This might seem like a scary amount of energy to unleash, but is actually rather little. 1TeV is about the same as the kinetic energy of a flying mosquito. What allows this to unravel the hadrons of matter is that the energy is concentrated into a space a trillion times smaller than a mosquito.

## Fig. 1 – Particle accelerator

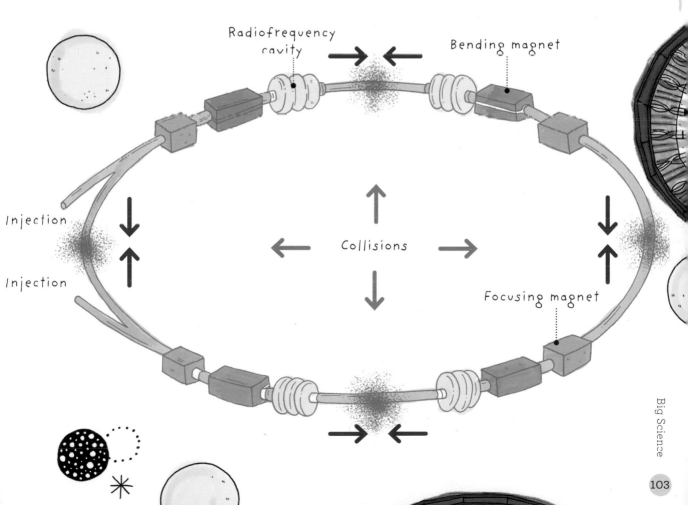

Radiofrequency cavity

Bending magnet

Injection

Injection

Collisions

Focusing magnet

# THE HUMAN GENOME PROJECT

Sometimes compared to the achievement of landing people on the Moon, the sequencing of the human genome is one of the great scientific triumphs of the age.

Francis Crick and James Watson (with some unacknowledged help from Rosalind Franklin) discovered in 1953 that the DNA molecule takes the form of a double helix. The strands are composed of phosphates and sugars, held together by the hydrogen bonds between pairs of nucleotide "bases" – adenine, thymine, guanine and cytosine, which couple up into base pairs, A with T, and G with C.

The task of the human genome project was to transcribe all the base pairs in a human, in the correct order. When the project started in 1990, nobody even knew how long this code would be.

This great feat of biology, carried out in laboratories in the USA, UK, France, Germany, Japan and China, relied on the physics of a process called electrophoresis.

The raw DNA, obtained from a small number of volunteers, first had to be cut into manageable sections. Every cell in your body contains 23 pairs of chromosomes, each a tightly-wound package of DNA. A combination of heat and chemicals unwinds the package, dividing it into shorter lengths, and making the double helix unzip down the middle.

Because the base nucleotides will only link up with their partner bases, the single strands can be repaired by introducing a supply of A, C, G and T. These molecules are picked up, resulting in a new double helix from each half of the original DNA. The process is repeated so that a single strand of DNA is multiplied many times, providing a large sample of many identical short sections of DNA.

Now the process is modified slightly, so that some of the new bases added to the mix are labeled with a fluorescent dye molecule. The half-strands of DNA again begin to assemble themselves back into double strands, but the process stops when a labeled base is added.

Instead of lots of identical copies of itself, the DNA has now cut itself into shorter sections of varying length, with a labeled molecule at the end.

The sample is sorted by the process of gel electrophoresis. The DNA molecules have a negative electrical charge, and are driven through a transparent gel by an electrical current. Shorter strands travel more quickly and so move farthest, and after a period of a few minutes the sample has been sorted into size order, each contained in a distinct band on the gel.

The labeled molecule at the end of the strand will fluoresce under UV light, revealing the identity of the base – A, C, G or T. As long as enough strands are tested, there will be samples cut at every point along the strand and all that is needed is to read them off in size order to be able to decode the entire sequence of that fragment.

Computers then perform the huge task of working out where all the sequences fit together and, eventually, create a complete sequence of code for a human being.

It's a huge task, as it turns out there are about three billion base pairs in the human genome. This means that if every letter in this book were replaced with the correct sequence of ACGTs, you would need more than 8,000 copies, occupying over 100m (328ft) of shelf space, to write it all down.

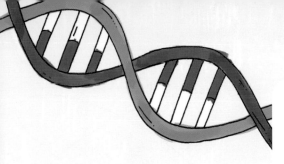

The first human genome took 13 years and cost about $3 billion to decode. Most DNA is identical from person to person, and it is possible to decode just the bits that vary, so today you can buy your own code of life for a few hundred dollars. It will fit on a DVD.

## Fig. 1 – How DNA fits into a chromosome

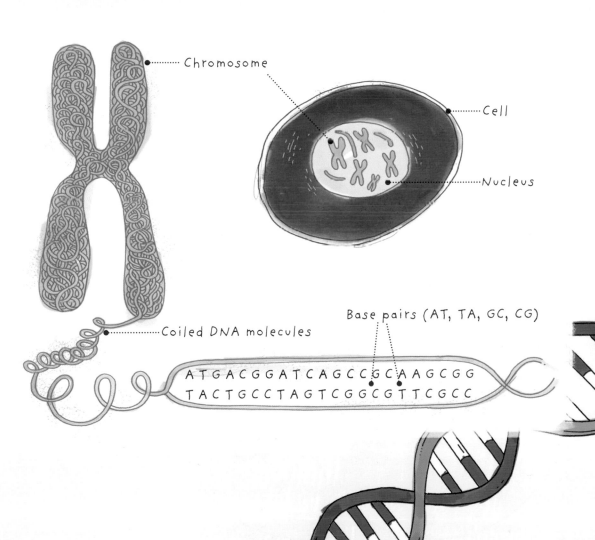

Chromosome

Cell

Nucleus

Coiled DNA molecules

Base pairs (AT, TA, GC, CG)

ATGACGGATCAGCCGCAAGCGG
TACTGCCTAGTCGGCGTTCGCC

# THE STANDARD MODEL

Almost everything you see around you, the matter that makes up your environment, is made of molecules which are in turn made of atoms. Atoms are the elements, the 118 (at last count) members of the periodic table.

**A**toms are made of three subatomic particles, the proton and neutron (together called nucleons because they form the nucleus) and a cloud of electrons. The number of protons in a nucleus determines which element it is, and the number of neutrons which isotope.

The nucleons are about the same size as each other (the neutron very slightly heavier), but both are massive compared to the electron. A proton is 1,836 times heavier than an electron.

In the early 1960s it was discovered that the nucleons were not fundamental particles, but were themselves made up of even tinier components. The electron, on the other hand, is one of the basic units of the Universe.

## What matter is made of

Within a decade the Standard Model was born, providing a sort of construction kit for matter. Two types of basic particles were found: the quarks and the leptons. Each comes in six varieties, which are paired into "generations".

One of the **leptons** is quite familiar – the electron. The others are the muon and tau, and each of these is paired with a neutrino variety: electron and electron neutrino, muon and muon neutrino, tau and tau neutrino.

The **quark** family names are even more esoteric: up and down, charm and strange, top and bottom.

## Fig. 1 – Particle zoo

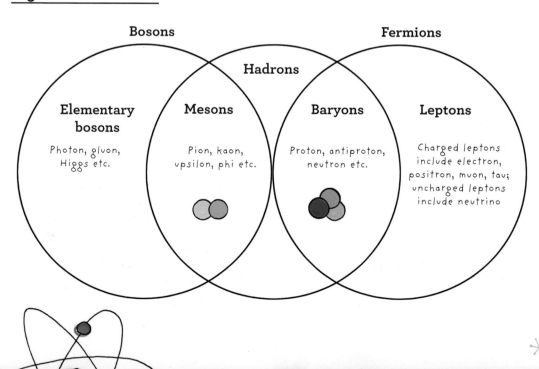

Bosons

Fermions

Hadrons

**Elementary bosons**

Photon, gluon, Higgs etc.

**Mesons**

Pion, kaon, upsilon, phi etc.

**Baryons**

Proton, antiproton, neutron etc.

**Leptons**

Charged leptons include electron, positron, muon, tau; uncharged leptons include neutrino

## Building blocks for the Universe

So far, so good. We have 12 basic building blocks and can use them to make things (nucleons) that are in turn used to make bigger things (atoms) that make the molecules that make the Universe. A neutron, for example, is made of one up and two down quarks, while a proton is two ups and one down.

But you can't just assemble your quarks any way you want. They come in three different "colours" which have nothing to do with the everyday meaning of colour but describe a type of charge or colour-force they carry. The colours are known as red, green and blue, and particles have to be assembled in such a way that they are colour-neutral. This does not mean they are beige.

Further levels of complication now arise. Each quark has a corresponding antiquark, which carries an anticolour. To achieve a stable "colourless" assembly, you could take three quarks of different colours (making a **baryon**), or a quark and an antiquark (making a **meson**). Mesons and baryons are together referred to as hadrons, a word familiar from the Large Hadron Collider (see page 102).

The quarks and leptons together are the **fermions**, and the Standard Model also includes a different class of particles called **bosons**. Bosons include photons and are the force-carrier particles implicated in the fundamental forces of the Universe.

## The force is strong

Four fundamental forces are known, and so far none of them respond to Jedi mind tricks. The most familiar is gravity, which holds our daily lives in place. Although it seems important, gravity is easily the weakest of the four.

The electromagnetic force is carried by photons, while the weak force is carried by W and Z bosons and the strong force by bosons called gluons (because they glue things such as atomic nuclei together).

What about gravity? This should be carried (or mediated, in the terminology) by a particle called a graviton. Nobody has yet found a graviton, which is one of the big problems with the Standard Model.

It's not the only one though. Physicists are still trying to reconcile aspects of the model with other theories and observations to create a truly universal model of the Universe.

## A duck with an Irish accent

**Physicist Murray Gell-Mann originally named the sub-subatomic particles he discovered "kworks", but found the spelling "quark" in the book *Finnegans Wake* by Irish author James Joyce. It has since been described as the sound made by ducks in a park in Dublin, Ireland.**

# GRAVITY

Gravity may be the most familiar of the natural forces in everyday life, but it is also the least understood. We don't really know what it is, how it works or where it fits into the various competing theories to explain the Universe.

The scientific study of gravity goes back to Isaac Newton, who famously observed an apple falling from a tree in his garden in Lincolnshire. He reckoned the force of gravity must be a property of matter, and that everything with mass carries a proportional force of gravity with it. This is a perfectly good model for everyday life, but doesn't quite cut it in terms of explaining what is really going on.

Albert Einstein offered a radically different view in his general theory of relativity. He said that mass distorts the geometry of spacetime, and this distortion is what draws objects toward each other. Gravity is a function of space, not matter.

## Gravity lens

Again, this works pretty well and explains things that were unknown in Newton's time. The most spectacular example, which was predicted by relativity before being observed, is the phenomenon of **gravitational lensing**. When two massive (we're talking galaxy scale) objects happen to line up from our viewpoint, light from the farther object can be bent around the closer one so that a good telescope can detect its light. The picture is distorted, broken into as many as four separate images which have traveled in different paths through the gravitational lens, but it means we can see around corners and get glimpses deeper into the cosmos.

Einstein's geometric spacetime distortion offers a good explanation of the way we see gravity working, but does next to nothing to explain why and how it happens.

## Fig. 1 – Global gravity variations

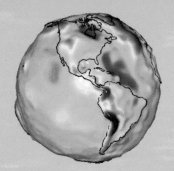
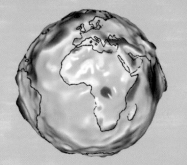
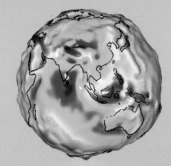

The difference between dark red (highest gravity) and dark blue (lowest) amounts to 100 milligals, or a variation in gravitational acceleration of about 10cm/s². For an average human this is a weight difference of 6g.

## Fig. 2 – The four forces

| Force | Carried by | Strength |
|---|---|---|
| Strong nuclear | Gluons | 1 |
| Weak nuclear | W+, W- and Z bosons | $\frac{1}{1,000,000}$ |
| Electromagnetic | Photons | $\frac{1}{137}$ |
| Gravity | Unknown | $\frac{6}{10^{39}}$ |

## How to lose weight

We've all experienced what seems to be a high gravity day from time to time, usually on a Monday morning. Why else would it be so hard to get out of bed? Have a little sympathy, then for people living in actual high-gravity zones.

From 2002 to 2017, a pair of spacecraft orbited the Earth to map variations in gravity in unprecedented detail. The GRACE mission (Gravity Recovery and Climate Experiment) identified zones of high and low gravity across the planet with unprecedented detail.

Good news if you live in the Midwest USA, Brazil or India: these are all low-gravity zones. Tough luck though if you are in the Pacific Northwest, Turkey or Indonesia. If you happen to be in one of these high-gravity zones, simply by moving to a low-gravity area you will weigh less. For an average human of 62kg (137lb), the difference is not much, but is measurable. Going from the highest to lowest gravity would amount to about 6g (about the weight of a teaspoon of sugar).

## Feel the force

There are only four known forces in nature: gravity, electromagnetism, and the weak and strong nuclear forces. By far the most powerful is the strong nuclear force, but this is only effective over a tiny distance – the diameter of an atomic nucleus. The weak nuclear force, as its name suggests, is much weaker. It is just one millionth as strong as the strong force, and effective over a much smaller range of a fraction of one percent of the diameter of a proton.

Neither of these two play any part in our normal everyday lives, except for holding together the atoms that make everything we can see and touch.

The electromagnetic force is much more pervasive, and manifests itself in everything from your microwave oven (see page 18) to the motor in your vacuum cleaner. It is less than one percent of the strength of the strong force, but vastly stronger than gravity. You may have a refrigerator magnet, perhaps a souvenir of a holiday. It may be tiny and not especially powerful, but it is able to exert enough electromagnetic force to overcome all the gravity of the planet Earth.

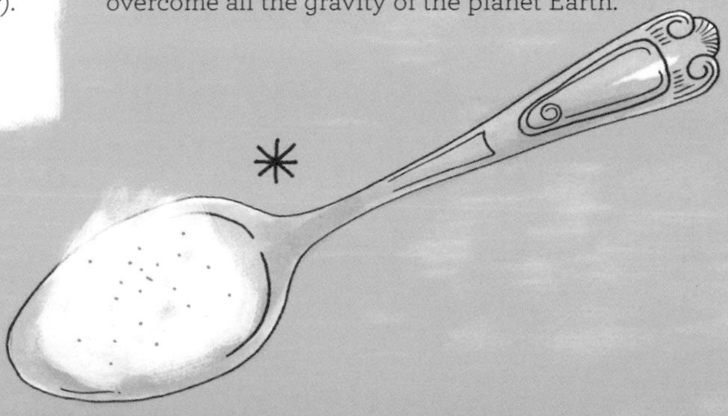

## Feeble mystery

The extreme feebleness of gravity is one of its big mysteries. It is 44 trillion trillion trillion ($4.4 \times 10^{-37}$) times weaker than the force that keeps your souvenir on your refrigerator.

Another mystery is just what causes it. The Standard Model of physics describes 17 fundamental particles, 12 of them fermions that make up matter, and 5 bosons which carry force (see page 106). A further seven bosons may exist, but have never been seen and are not accounted for in all models. Conspicuous among the "missing" bosons is anything that might be responsible for gravity.

The most recently discovered boson is called the Higgs, after Peter Higgs who predicted its existence in the 1960s. It was finally detected by the Large Hadron Collider (see page 102) in 2012. The Higgs generates a field that imparts mass to any fermion passing through. Without mass there would be nothing for gravity to work on, but the Higgs does not carry gravitational force itself.

If there is a gravity-carrying boson, the hypothetical graviton, it has never been found or even hinted at. The mathematics suggest that it would be practically impossible to build a graviton detector, given that they are so feeble, with any conceivable technology.

## Ripples in spacetime

But that hasn't prevented the detection of **gravity waves.** One of the most important discoveries of our time is that of the ripples in spacetime caused by the collisions of black holes and neutron stars deep in space.

Gravity waves are unbelievably small. The detector that first found them, known as LIGO (Laser Interferometer Gravitational-Wave Observatory) consists of two huge machines 3,000km (1,864 miles) apart in Louisiana and Washington State, USA. Each has a pair of tunnels 4km (2.5 miles) long, arranged at right angles, along which a powerful 200W laser is fired.

By reflecting the laser 280 times up and down each tunnel, the effective length of the beam was increased to 1,120km (696 miles). The idea is that, if a gravity wave were to pass the detector, it would stretch the beam in one tunnel by a larger amount than the other – and this is exactly what happened.

The difference measured is hard to comprehend, but it was real. Two vast black holes spiralling to their mutual annihilation millions of light years from Earth generated a ripple in the fabric of the Universe which, when it reached us, amounted to just one thousandth of the width of a proton.

## Fig. 1 – Colliding neutron stars generate measurable gravity waves

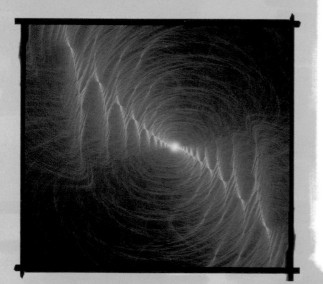

## The last mystery

There is one further thing about gravity that makes it different from the other three basic forces: it only pulls. The other forces can act as attractive or repulsive forces (two magnets can attract or repel, for instance), so why not gravity?

Being able to take antigravity out of science fiction and into everyday reality would be very exciting. To do that, the most likely process would be to discover antigravitons and then find a way of controlling them. As the (positive) graviton is still nothing more than a theoretical possibility, the prospects for antigraviton-powered flying cars look poor – at least in the near future. More research needed!

# EVERYTHING

Understanding the physics behind everything is the modest ambition of several groups of scientists who are trying to find a way to join the dots between the two great theories.

**W**e have a really good model of how the Universe behaves at the large scale. This is Einstein's **theory of relativity**, and it provides lots of answers about the way stars, galaxies and even human-scale objects like satellites and aircraft behave. Relativity is entirely consistent with the more intuitive everyday mechanical physics of Isaac Newton. At low speeds and smaller dimensions, Newtonian physics describes things perfectly. At higher speeds and over cosmic scales, relativity fills in the gaps and explains apparent anomalies.

We also have another really good model, describing how things work at very small scales. **Quantum field theory** provides verifiable predictions and good explanations of many phenomena that appear to defy common sense. Nobody will pretend that quantum physics is as easy to absorb as Newtonian mechanics though. It is sometimes said that if you "get" quantum mechanics, you haven't really understood it at all. But it remains the best working explanation of events at scales smaller than an atom.

It seems only right that there should be a single set of rules for the Universe, which could be applied to any object or event at any scale, but relativity and quantum theory just will not agree with each other. They can't even agree on the basic forces of the Universe.

## Use the force, look

Relativity explains gravity as a distortion in the geometry of spacetime caused by the presence of matter. This distortion is what keeps objects in orbit, and even predicts some quite startling phenomena.

The warping of spacetime should cause light from distant objects to be bent around massive objects, so that the image appears to be in the "wrong" place. This gravitational lensing effect was predicted before it was observed, but has now been seen many times. This might seem to seal the deal for relativity.

Another argument in favour of relativity is that gravity doesn't fit well in quantum theory. Each of the other forces – electromagnetism and the strong and weak nuclear forces – are carried by subatomic particles called bosons. The weak force is carried by W and Z bosons, and the strong force by gluons (the glue that holds a nucleus together). Light is carried by photons, another of the boson family, and the Higgs boson, which imparts mass to objects, was one of the great discoveries of the Large Hadron Collider (see page 102). Among other things, this discovery helps unify the weak and electromagnetic forces into a single "electroweak" force.

Quantum theorists are hunting for a boson to carry gravity, the graviton, but despite all their efforts it has never been seen. This is strange because, if it exists, current technology should have revealed it. Even the discovery of gravity waves has done nothing to track down the elusive and still-theoretical graviton.

Yet quantum theory answers many questions that relativity cannot – radioactivity, for example. This clearly demonstrates the quantum nature of the subatomic universe – that there are packets of energy (quanta) that are the smallest possible, and cannot be divided.

## One rule to rule them all

The basic incompatibility of the two great models of the Universe makes physicists uncomfortable, because it seems that something really important has been missed. We can't simply have one set of rules for one set of circumstances, and another for another. What would happen in a situation where both circumstances applied?

A case in point is a black hole. This has huge gravity (relativity) but tiny size (quantum). What set of laws would work here?

Attempts to create a single universal theory of everything – a grand unified theory or unified field theory – have led to some quite bizarre-seeming ideas. String and superstring theory and supersymmetry (see Strings, rings, and other weird things, page 116) invoke many new physical dimensions and have produced advanced logical arguments but no experimental evidence.

> Radioactivity demonstrates that there are packets of energy that cannot be divided. This can only be explained by quantum theory.

And this is the nub of the problem. We can conduct experiments that "prove" relativity, and others that "prove" quantum theory. It is hard to disprove a theory, but if you look for expected evidence and it consistently fails to materialize, eventually you must conclude that the theory is wrong.

Quantum theory predicts that protons should decay. The half-life of a proton should be about 23 orders of magnitude longer than the life of the Universe, and statistically it should have been possible to observe such a proton decay. To date there is no evidence of this.

> Theoretical half-life of proton = $10^{32}$ years
>
> Age of the Universe = $1.4 \times 10^{9}$ years

None of this means that a theory of everything is impossible, but it looks like neither relativity nor quantum theory will ever claim that status.

## Fig. 1 – Radioactive decay in uranium-238

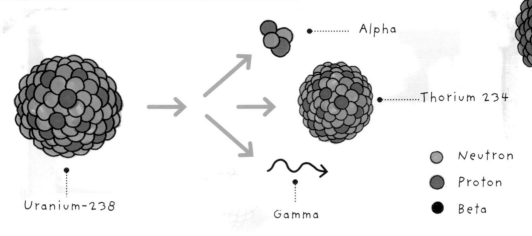

## End of the line, again

It is thought that such a theory would rapidly put physicists out of work, and mean an end to science. After all, once everything has been explained, all that's left is a little mopping up and measurement.

Physics has been here before, more than once. Legend has it that Lord Kelvin, president of the Royal Society and one of the most eminent scientists of the 19th century, said at a meeting of the British Association for the Advancement of Science in 1900: "There is nothing new to be discovered in physics now." Whether he said it or not, it reflects the thoughts of many at the time. Just five years later Albert Einstein discovered something new – special relativity.

One of the founders of quantum mechanics, Max Born, is quoted in 1927 as claiming that "physics, as we know it, will be over in six months". Further claims that the end was in sight have been made, often by eminent scientists, for the succeeding century. And every time, the goal of a true theory of everything has remained tantalizingly just out of reach.

## Universal and eternal?

Among the questions that need to be asked when thinking about any theory of everything is whether there really is just one set of laws for the whole Universe.

Distant galaxies and quasars, the orbits of black holes and nebulae, all do seem to follow the same rules. No phenomena anywhere in the observable Universe has been found that breaks this.

But it is interesting to note that, in Isaac Newton's time, a "clockwork" Universe seemed an almost perfect model. With Einstein, a rather more complex model was also nearly perfect. Quantum field theory gives a still more complicated and also nearly perfect model. It's almost as if the laws of the Universe are not constant, but evolving in complexity to stay always one step ahead of our understanding of them.

WEIRD UNIVERSE

# STRINGS, RINGS AND OTHER THINGS

Before there was proof of the existence of atoms, it was not known whether matter could be divided into ever-smaller pieces without limit. By the early 19th century it was realized that the atom was the basic unit of each chemical element, and before long the main subatomic particles were determined. The universal arrangement of a nucleus containing protons and (usually) neutrons, with a cloud of orbiting electrons, became understood.

Today we've delved farther down the well, and discerned most of the building blocks of the main subatomic particles – the quarks, leptons and bosons of the Standard Model. But what are *they* made of?

It's convenient to think of all of these particles, from the proton downward, as tiny balls of stuff that bounce around a bit like billiard balls. Unfortunately, this simple model doesn't really work even at the comparatively huge scale of the proton and neutron, and certainly not once we're investigating electrons and other apparently elementary particles.

Part of the weirdness of quantum physics is that the idea of an electron whizzing around the nucleus, in a similar way to a satellite orbiting the Earth, has to be abandoned. Depending on how you look at it, an electron can be either a particle or a wave, and exists not at a specific location but as a probability function somewhere within the cloud. If electrons and their cousins have a tangible existence, they can be considered to exist at a practically dimensionless point.

So far, so good(ish). The Standard Model works pretty well and is mostly consistent with itself, but doesn't quite answer all the questions. One prominent theory that attempts to reconcile these and account for the actual nature of the Universe is **string theory**.

## Pick your string (theory)

One of the least confusing things about string theory is that there isn't just one of them, but a number of alternative versions. In conventional string theory, what we perceive as the fundamental units in the particle zoo are actually the manifestation of vibrations in strings that exist in dimensions beyond our regular 3D or 4D Universe.

Strings are one-dimensional, and appear as points where they intersect our three physical dimensions. The nature of their vibration determines how they appear to us – as electrons or photons or blue charm quarks.

Some versions of string theory hold that the strings are linear, with a beginning and an end, while others consider that they are small closed loops. In either case, the mathematical acrobatics necessary to make this conceivable means that there need to be many more than the three physical dimensions we know in everyday life.

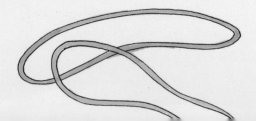

# Super particles

Standard string theory suggests that there are ten (or possibly eleven) physical dimensions in the Universe, while an early version known as bosonic string theory needed as many as twenty-six.

Superstring theory is one of the current front runners of the many competing schemes, and is closely tied into the concept of **supersymmetry**. This predicts that every particle in the Standard Model, including the force-transmitting bosons (photons, gluons, the Higgs and others) and the constituents of matter, the fermions (electrons, neutrinos, quarks and so on), all have a corresponding "superpartner".

These theoretical particles all need names, and the convention adopted is to add "ino" to the end of a boson name, or an "s" to the beginning of a fermion. This leads to some delightful new monikers: the superpartner of an electron is a selectron, and of a quark is a squark. A photon has a corresponding photino, while the boson known simply as the W has a superpartner called a wino.

There is one more contender that demands a mention. M-theory attempts to reconcile the five competing versions of superstring theory, introducing the notion of "branes" (membranes) of which the strings (or loops, or superstrings) are a manifestation. The M might stand for membrane, but nobody seems entirely certain about this and it might as well stand for magic or mystery.

Some of the greatest branes – sorry, brains – in theoretical physics are trying to make these models fit the Universe as we see it, or perhaps make the Universe fit the model. It's enough to make a wino of a teetotaller.

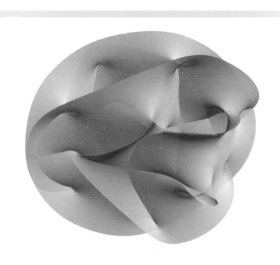

An attempt to depict a multi-dimensional brane as a three-dimensional image on two-dimensional paper

# N-DIMENSIONAL SPACE

The Universe we experience on a daily basis has three physical dimensions and one dimension of time. If string theory, superstring theory, M-theory or supersymmetry is true, the Universe must actually contain at least ten or eleven physical dimensions. Where are they, and why can't we perceive them?

It is hard to imagine any extra physical dimensions, but it can help to take our thoughts in the other direction and imagine one fewer. Let's say we exist on a two-dimensional sheet or membrane, which itself exists within the three-dimensional world we know. The membrane presents no sort of barrier to outside objects, but some strange force means that we can only move in two dimensions – left/right, and forward/backward.

What's more, our eyes can only detect objects in this 2D world, and see nothing above or below. There is, as far as we are concerned, no such thing as above or below.

Any object, whether it was another creature like ourselves or some inert lump of something, would appear to us as one-dimensional. A line. In our 3D world we see 2D images of things and understand they also have depth, and in this 2D flat universe we can use imagination to understand the second dimension.

Your entire existence has been in this plane, making it very difficult for anyone to explain where the third dimension is.

## Entering a higher dimension

Now, what if a 3D object from our Universe passed through this flat realm? Let's say a simple shape such as a ball.

In this 2D universe we don't understand the idea of a "ball". What we see is an object which appears, at first just as a point but rapidly growing in size to become a substantial 1D line. We can understand that it has depth, and indeed we might discuss with our friends how this line seems to be the same width from every viewing point. It wouldn't be hard to work out that it was in fact a circle.

But this circle is changing size in an unusual way. It appeared as if from nowhere, grew bigger and then, as it continued its journey through our world, began to shrink again. Before long it disappeared altogether, leaving no sign it had ever existed. This thought experiment might give a little insight to the way an n-dimensional universe might interact with our own perceived 3D one. Objects from other dimensions may pass through our world, appearing as if from nowhere, morphing in some strange way, and then disappearing again.

Do we see anything like this in real life? Yes, we do.

# The thing from beyond space

One answer that is more metaphysics than physics, is to think of life itself as an interaction from another dimension: it starts from a tiny dot, grows, shrinks, and disappears.

On a more scientific basis, natural quantum fluctuations are understood to allow fundamental particles to spontaneously come into existence. The Big Bang theory is often thought of as an explanation for the creation of the Universe, but really it tells us what seems to have happened from the point of creation to the present. It explains that the Universe grew from a singularity – an infinitesimal zero-dimensional singularity containing everything that we consider to exist – through a process

of expansion, to the Universe we see today. It does not quite explain how that singularity came into existence, other than suggesting this happened spontaneously.

Our Universe, then, came from nothing, and has grown. Common sense would expect that the growth would slow down after the initial burst, but in fact it is accelerating (common sense is proving to be a not very valuable commodity in theoretical physics).

It's almost as if the whole Universe is actually the interaction of a thing from other dimensions passing through our own otherwise-empty four-dimensional spacetime.

## Fig. 1 – Basic spatial dimensions

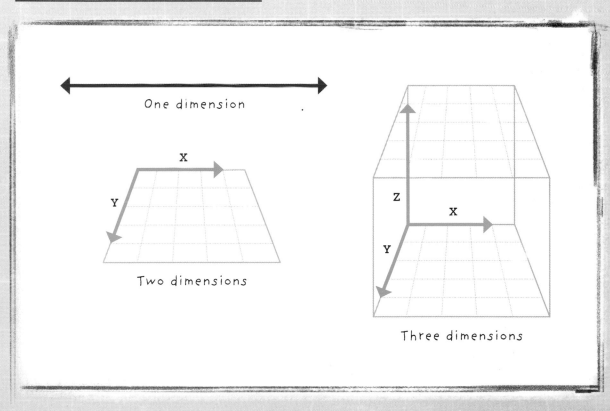

One dimension

Two dimensions

Three dimensions

# THE HYPERCUBE

What on Earth is a hypercube? Rather like a cube, which exists in three dimensions, a hypercube is a solid object – but occupying four spatial dimensions. You aren't likely to stumble across one in your everyday life, but the mental exercise to even conceive of one is useful when trying to grasp the strange possibilities presented by multi-dimensional spacetime.

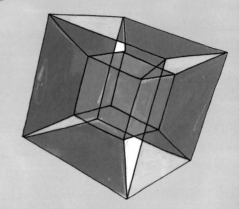

**W**e are familiar with three everyday dimensions of space: length, breadth and height. We tend to think of time as a fourth dimension, but for the purposes of the hypercube we're going to put time to one side and try to imagine a fourth physical dimension.

Let's work our way up.

A point occupies no dimensions. It's just a dot of zero size, fixed by its position in whatever coordinate system we are using.

Join two points together and we have a line – a one-dimensional artefact. Four equal lines meeting at right angles is a square, occupying two dimensions. Next, we can assemble six squares to form a three-dimensional cube. Now we're getting somewhere.

Each stage involves using a number of components from the next lowest number of dimensions to create an object occupying a higher number. How do we go further?

A hypercube would be a four-dimensional object composed of eight cubes. You would build it by assembling regular 3D cubes on each face of a cube, in such a way that everything remains perpendicular to everything else.

But hang on – eight cubes? How can we assemble eight cubes when there are only six faces on a 3D cube? Let's go back a step: a 3D cube is made by assembling together six squares, on a base square having only four sides. We do it by moving into another physical dimension. All we have to do is arrange our eight cubes in a similar fashion.

## Ceci n'est pas une hypercube

It seems impossible, and in practical terms it is. We might have more luck making an image of one.

We can easily create an image of a 3D cube on 2D paper. It makes sense then that we can create an image of a 4D object on a 3D surface. A 2D drawing of a 3D cube on each face of a cube is a first approximation. Better would be to put a hologram – a 3D image contained within a 2D surface – on each face.

The object we have created is getting closer to a representation of a hypercube. We still only have six holographic cubes, although we might consider a seventh in the heart of the assembly and an eighth enveloping it. The temptation is to find "space" for them all in our conventional 3D framework of everyday life, but remember the point of the exercise is to try to imagine a 4D framework.

If this is difficult, try moving up a further dimension. How about a 5D cube? Or to really blow your mind, forget the straight lines and right angles: can you imagine a 4D hypersphere? No, me neither.

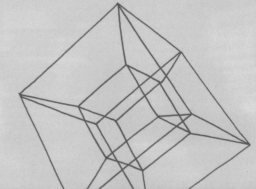

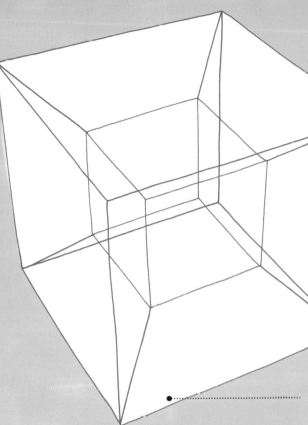

## Square seconds and cm⁴

Moving beyond three dimensions in geometry is an interesting diversion, but the use of higher-dimensional units is standard practice in some aspects of everyday life.

Acceleration is measured in metres per second per second, or $m/s^2$. Time squared! The rate of change of acceleration is seconds cubed.

In structural engineering, a property known as the moment of inertia (or second moment of area) is used to work out how much a beam will deflect under load. It is measured in units of length to the power four, often $mm^4$ or $cm^4$, which is mathematically what a hypercube would be.

A common representation of a hypercube, also called a tesseract, looks something like this. It's hard to represent four dimensions on a 2D sheet of paper.

## Fig. 1 – Moment of inertia

These identical beams have different moments of inertia, calculated in $cm^4$

# ANTIMATTER

The stuff of science fiction plots, antimatter is as real as ordinary matter. It can be created and harnessed as a matter of routine, yet it presents some fundamental mysteries. Chief among these is why there seems to be so much more matter than antimatter in the Universe.

**E**ach fundamental particle has a mirror-image antiparticle – the same mass but with opposite charge and spin. Electrons have a negative charge, while their antiparticles, the positrons, have positive charge. Protons are mirrored by antiprotons, and neutrons by the antineutron. See The standard model, page 106 for a full breakdown.

It might seem odd that neutral charge can have an opposite. Neutrons are made of three quarks (see N-dimensional space, page 118), and each quark has an antiquark equivalent. All you need is the right three antiquarks and, hey presto, you have an antineutron.

At the Big Bang, particles and anti-particles should have been created in equal numbers. Particle–antiparticle pairs sprang into existence, and quite often sprang straight back out again. When a particle meets its antiparticle, the two destroy each other in mutual annihilation, leaving behind just energy in the form of photons.

The process also works in reverse – when energy is destroyed it leaves a residue of matter and **antimatter**. With all the collisions going on in the very early Universe (we're talking about the first fractions of a second of existence), most of the stuff that came into being should have disappeared almost immediately in a puff of energy. What's left would be either energy – if all the matter and antimatter got destroyed – or equal amounts of matter and antimatter.

We can easily see that all the stuff was not annihilated, otherwise we wouldn't be here to wonder about it. The mystery remains: The Universe today is made almost entirely of matter, with hardly any antimatter. And nobody is quite sure why.

It seems for every billion early matter–antimatter collisions, one particle of matter escaped.

All of the great particle accelerators on Earth have produced only minuscule amounts of antimatter. Fermilab, the Large Hadron Collider and DESY together have created less than 20 nanograms of the stuff, making it hard to study. We don't even know for sure whether gravity attracts or repels antimatter.

Among the possible explanations for all this are that the laws of physics do not apply equally to antimatter, or that elsewhere in the Universe there are places where matter is the rarity and antimatter the norm.

## Everyday antimatter

It's exotic stuff, antimatter, but it is around us every day. If you've ever had a **PET scan** (positron emission tomography), you have had antielectrons released into your body from a radioactive tracer. When these positrons find an electron, the matter–antimatter collisions result in them being destroyed, creating pairs of photons which can be detected to produce useful medical images.

You don't even have to go to a hospital to be in the presence of antimatter. Thunderstorms produce beams of antimatter particles during a terrestrial gamma-ray flash, when lightning is created. This happens hundreds of times every day, but you might be relieved to know that the antimatter beams are directed away from Earth into space.

## Fig. 1 – Antiparticles

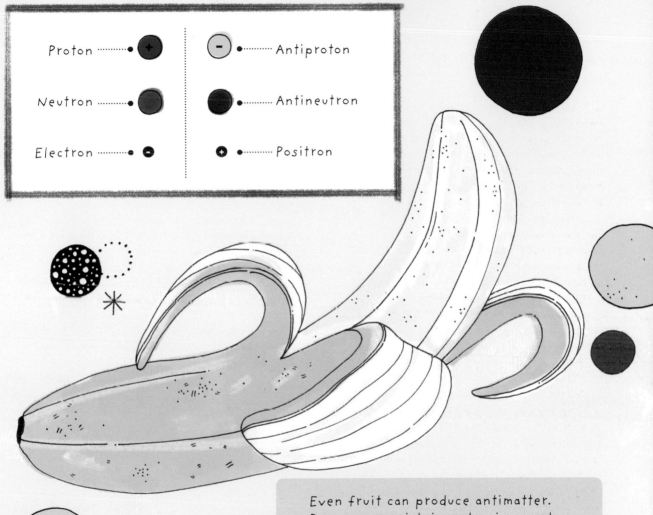

Proton ........ • **+**        **−** • ........ Antiproton

Neutron ........ • ●        ● • ........ Antineutron

Electron ........ • **−**        **+** • ........ Positron

Even fruit can produce antimatter. Bananas are rich in potassium, and a small fraction of this is the isotope potassium-40. This naturally decays via positron emission (just like in a PET scanner), so that an average banana generates a particle of antimatter about every 75 minutes.

# THE DARK UNIVERSE

There is a long-standing problem in cosmology, the branch of physics that deals with really big issues such as how the Universe is structured and where it came from: most of the Universe is missing.

Imagine you see a litre bottle of water: you would be fairly confident that it weighed about a kilogram. If you picked it up and found it only weighed 50g, you might think you had been tricked somehow.

This is essentially the trick the cosmos has been playing on us. By observing the motions of galaxies and quasars, and the bending of light due to the curvature of spacetime, it is possible to deduce how much gravity is pulling on them. And if you know how much gravity there is, you can work out how much matter there must be.

You then tot up all the matter you can see. Add together all the stars and nebulae and dust clouds and swarms of comets and asteroids and everything else. The total is a small fraction of what is expected. Most of the matter is missing.

## Lost and found

Part of the problem was solved in late 2017 when two groups of scientists independently announced that they had found the missing matter. It was hiding in plain sight, strung out in tenuous "filaments" of gas stretching between galaxies.

The material evaded detection by being at an awkward temperature – too hot to be seen by its absorption of background radiation, and too cold to emit its own. It's incredibly diffuse, amounting to just a handful of atoms per cubic metre, but about five times more dense than regular intergalactic space.

Problem solved? Well, not quite. What was found was the missing baryonic matter of the Universe – ordinary matter as you and I know it. Baryons are the protons and neutrons and other not-very-exotic particles of everyday existence. But this only accounts for about five percent of the Universe, while mysterious **dark matter** accounts for a further twenty-two per cent.

Nobody is quite sure what dark matter is, because nobody has yet actually detected it. Galaxies are held together by the combined gravity of everything they contain, yet there just isn't enough stuff to do the job.

The amount of dark matter that should exist has been predicted from studies of the cosmic microwave background – the pattern of feeble radiation left behind by the Big Bang.

There should be between four and five times as much dark matter as baryonic matter. It must have mass, and it interacts very weakly with normal matter (if at all – the first confirmed detection of dark matter will certainly justify a Nobel Prize). Cosmologists therefore conclude that dark matter is made of weakly interacting massive particles, or wimps. And that's about all we know about them.

There is one further big mystery, which is a bigger problem than the previously missing baryonic matter and invisible dark matter put together.

The Big Bang threw all of the stuff of the Universe outward in an expanding bubble. Gravity puts on the brakes, so it is obvious that the expansion will be slowing down – it just makes sense.

In fact, the expansion is accelerating. This was such a profound shock to cosmology, when first discovered in the late 1990s, that a new explanation had to be found. Some unknown effect is acting contrary to gravity. This **dark energy** accounts for three-quarters of the Universe. Among its mysteries is that it is evenly distributed in space and time, so is not diluted as the Universe expands.

## Recipe for a universe

**Take 73 percent mysterious dark energy, 22 percent mysterious dark matter, 5 percent stuff we can see, only half of which is mysterious. Bring to the boil and simmer for a few billion years.**

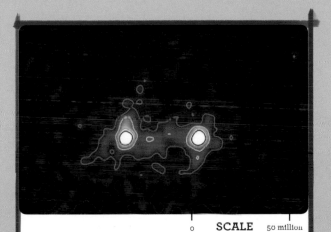

| 0 | SCALE | 50 million |

Captured by astronomer Mike Hudson at the University of Waterloo, Canada, in 2017, this is said to be the first ever image of dark matter — found in filaments (red) between two galaxies (white).

# QUANTUM WEIRDNESS

Quantum physics can be described as the study of things that just don't make sense. It mostly concerns very tiny phenomena including subatomic particles and discrete units of energy such as the photon. At this scale, the common sense rules that govern physics in the macroscopic (real) world just don't apply.

One of the most familiar of many quantum paradoxes is demonstrated by the double slit experiment. Light is shone through two slits in a barrier and onto a screen behind. Light passing through each slit will encounter light passing through the other, and the two will interact. In some places the energy of the two will add together, producing a brighter light, and in other places it will cancel out. This **interference pattern** is clear evidence of the wave nature of light.

If this is repeated with particles of matter – atoms or electrons, say, you might expect the pattern on the screen to be much simpler. Particles of matter should behave like grains of sand and simply pass through the slits to accumulate in heaps on the other side. But they don't. The same sort of interference pattern is created whether the beams passing through the slits are energy (photons) or matter.

It appears that small packets of matter passing through the slits somehow interfere with each other, changing their trajectory to produce the pattern.

This is already a bit odd, but it soon becomes downright bizarre. If you fire the particles one at a time, there should be no way for them to interact. Each will pass through one or other slit (or bounce off the barrier), making a mark on the screen before the next one stars its journey. Shouldn't this produce a simple two-heap pattern?

In fact, the same interference pattern emerges. The tiny lumps of matter are behaving just like waves, somehow arranging between themselves to build up a pattern on the screen. Do they pass through one slit, or the other, or somehow both?

Let's try to find out. Let's put a sensor on one slit to detect a particle passing through, and then we can perhaps make sense of it.

The unexpected result is that the particles now behave just like grains of sand, forming neat piles of impacts on the screen. But turn the sensor off, and they immediately create the interference pattern again. They seem to know when they are being watched, triggering the wave-like function to collapse in favour of particle-like behaviour.

This particle/wave duality is a core discovery of quantum mechanics, and the fact that simply observing can change the outcome of an experiment is central to many quantum paradoxes.

> Particles behave like grains of sand when they are being watched, but like waves when not.

# Fig. 1 – Double slit experiment

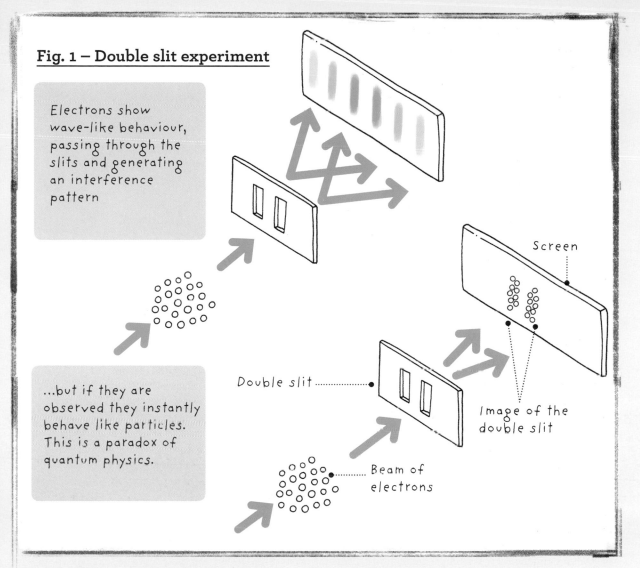

Electrons show wave-like behaviour, passing through the slits and generating an interference pattern

...but if they are observed they instantly behave like particles. This is a paradox of quantum physics.

Double slit

Beam of electrons

Screen

Image of the double slit

## Particle, wave, or both?

Pilot wave theory is an attempt to explain duality, suggesting that the tiny dots of matter (quanta) are not either a wave or particle, but consist of both at the same time. A wave component acts like a wave should, passing through both slits and generating a pattern of possible routes for the particle component – the wave acts as a pilot.

This gets around the bizarre nature of some quantum mechanics, and would mean that the Universe is essentially deterministic rather than probabilistic. A particle has a definite location at all times, and its trajectory could be predicted if its position and motion could be measured accurately. It also offers a possible explanation for quantum entanglement, in which two particles some distance apart seem to have knowledge of what the other is doing.

However, pilot wave theory doesn't answer everything, and even brings some new problems of its own. It doesn't explain the oddness of an observation changing an outcome, nor does it account for relativity. It's fair to say that some physicists like the simplifications it offers, while others hate the new complications that come with it.

# QUANTUM BIOLOGY

The meaning of life has perplexed great thinkers since the dawn of consciousness, and properly belongs within the realm of philosophy known as metaphysics. The *explanation* of life, however, owes more to actual physics.

Physicists like to argue that all science is physics anyway. Chemistry is underpinned by reactions at atomic and subatomic levels, and biology is a specialist sub-branch of organic chemistry with the added zest of life. This much is pretty standard "classical" physics, involving predictable events at almost-observable scales.

There is a growing volume of evidence that many key features of biology actually have more to do with quantum physics than classical physics. Quantum physics is where all common sense stops working, where it's normal to be baffled and weirdness comes as standard.

In quantum physics, an electron is both a particle and a wave at the same time, and can be in several places at once. The simple act of observing an experiment can change its outcome.

A biologist will tell you that enzymes cause chemical reactions to break or join biological molecules. That is certainly what they do, but not how they do it. It is now known that enzymes are able to use the bizarre effect of quantum tunnelling to transfer subatomic particles – electrons and protons – across cell barriers.

The sense of smell may be another quantum mechanical effect. Classic biology tells us that odour molecules are identified purely by their shape, which fit exactly into receptors in the nose and generate a signal the brain can interpret. There are billions of different smells that even we olfactorily challenged humans can detect, using just 400 different smell receptors.

However, the shape–fit (or lock and key) concept falls apart when it is realized that different molecules of the same shape are recognized as different smells, while molecules of different shapes but similar chemical structures give the same smell. Molecules of sulphur and hydrogen can take on many shapes, but all smell of rotten eggs.

The quantum mechanical explanation, known as the vibration theory of olfaction, is that the vibrations in the atomic bonds of a molecule are registered in the receptors, triggering the nerve signal. In effect, your nose is listening to the quantum vibrations of the odour.

Quantum effects have also been found in animal navigation (see Migration, page 144) and in the process that ultimately powers all life on Earth – photosynthesis. Conventional biology tells us that chlorophyll in green leaves uses sunlight to convert carbon dioxide and water into sugars. How it does this is unexpectedly strange. Bacteria inside the cells capture the photon (a quantum of energy) and deliver it to the reaction centre where it becomes the chemical energy of the sugars. It doesn't take it straight there, but sends it via multiple routes simultaneously – quantum coherence. The photon has adopted a "superposition" – being in several places at once.

In the famous Schrödinger's cat thought experiment, the cat is in a superposition (both alive and dead simultaneously) until the box is opened.

> Photosynthesis is now known to involve quantum coherence, in which the photons of light energy adopt superpositions.

## The ultimate question

Fans of *The Hitchhikers Guide to the Galaxy* will recall that a species of hyper-intelligent pan-dimensional beings built a supercomputer to answer the ultimate question of life, the Universe and everything. The gag was that the answer – 42 – was useless because the beings hadn't really understood the question.

Why 42? Douglas Adams, creator of *The Hitchhiker's Guide to the Galaxy*, was an avid computer enthusiast. In programming, the asterisk character * is used as a sort of wildcard, for a value that can be anything. And in ASCII code, the * character is assigned a numerical value of... 42.

The meaning of life? Adams was saying it is "anything you want it to be".

Reaction centre  Photon

Chlorophyll molecule

A photon striking a green leaf is delivered to the reaction centre in the cell — where its energy is used to make sugars — by multiple paths. This is one of many unlikely-seeming quantum phenomena in everyday biology.

Weird Universe

# TIME CRYSTALS AND MAJORANA

There is a lot of competition for the title of the weirdest of all, but for my money there are two front runners – time crystals and Majorana particles.

**T**hey sound like something out of *Doctor Who*, and seem to break fundamental laws of physics by demonstrating **perpetual motion** (which is, of course, impossible, see Perpetual motion heat death, page 180), yet time crystals are a thing.

Normal 3D crystals – diamond, quartz, table salt – have their atoms arranged in a regular three-dimensional lattice. So-called 2D materials such as graphene and boron nitride have their atoms arranged in flat sheets (they are called 2D but also have a very small third dimension, the height of an atom).

Time crystals are four-dimensional, with their atoms arranged in a constantly shifting pattern which repeats over time. They represent an entirely new phase of matter, in addition to the familiar solid/liquid/gas phases of everyday life and the extreme phases such as plasma and Bose-Einstein condensate (where extremely cold atoms clump together and act as one).

In a time crystal, ions (charged atoms) are in constant motion and repeat their arrangement after a period of time. Quite what that time is depends on the circumstances. In experiments, rings of atoms with entangled electrons are seen to flip with regular frequency. Stimulated by a pulsed laser, they flip with a frequency an exact multiple of the pulse frequency, but never at the same frequency.

With the laser turned off, the flipping continues despite there being no energy added to the system.

This looks just like perpetual motion, in contravention of the second law of thermodynamics, but can be explained because the crystal is a closed system with no energy extracted or added.

Besides being one of the oddest, time crystals are also one of the newest things in physics. They were first suggested as recently as 2012, and observed only in 2017.

## The self-antiparticle

Every matter particle is thought to have its corresponding **antiparticle**, each a tiny scrap of antimatter (see Antimatter, page 122). Just to make this slightly more strange, there is a class of particle known as Majorana fermions that are their own antiparticle. This suggests they are both matter and antimatter at the same time.

They are named after the physicist Ettore Majorana, who predicted their existence in 1937. The particles were first observed in 2012 but remain mysterious (rather like Ettore himself, whose unexplained disappearance in 1938 has sparked numerous theories).

Majorana particles can have no charge, ruling out protons, electrons and quarks. Antineutrons have been shown to exist, so they cannot be Majorana, but there is some intriguing evidence that neutrinos might be.

Neutrinos have almost no mass, which makes them very hard to study and earns them the nickname ghost particles. As you read this, billions are streaming through your body without having any effect on anything. Countless numbers pass right through the Earth every day, and only rarely does one make direct contact with another particle.

When a proton or neutron undergoes radioactive decay, it emits a neutrino and an electron or positron (antielectron). A double decay is when two particles do this together, and (in theory at least) the neutrinos could immediately annihilate each other leaving only the electrons or positrons (beta particles) and photons to be detected. If this neutrinoless double beta decay could be observed, it would confirm the Majorana properties of the neutrino.

In other experiments, Majorana particles have been seen to emerge from the ends of semiconductor wires in contact with a superconductor.

## Towards hypercomputers

Sometimes the realm of advanced physics can seem a long way from reality and of no practical benefit at all. Time crystals and Majorana fermions look like good examples of this, but they have a surprising and very exciting potential application in common.

Both are suggested as the basis of the quantum bits (qubits) that will make the quantum computers of the future, and which promise to make today's supercomputers look as basic as an abacus.

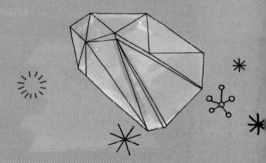

## Fig. 1 – Are neutrinos their own antiparticle?

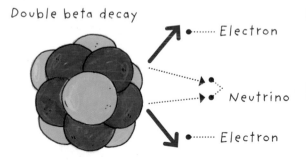

Double beta decay

Electron
Neutrino
Electron

Two neutrons convert into two protons emitting two electrons and two neutrinos.

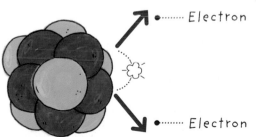

Neutrinoless double beta decay

Electron
Electron

Two neutrons convert into two protons emitting two electrons and two neutrinos which annihilate one another.

# THE ENVIRONMENT

# WEATHER FORECASTS

You have to have a little sympathy for weather forecasters. The atmosphere of planet Earth is in constant turmoil, driven by a number of different forces which, although quite well understood, are not necessarily predictable.

The basic mechanics of the climate are simple enough. It's warmer near the equator than at northern and southern latitudes, so the air heats up, expands and rises. This reduces the local pressure, so cooler air rushes in from north and south to even things out. The expanding warm air is drawn toward the poles to replace the cooler air which is now moving to the equator.

The simple **convection current** of warm air moving away from the tropics at high level, and cool air moving toward the tropics at low level, is set. But whatever the weather is, it certainly isn't simple. The size of the planet and the relative thinness of the atmosphere breaks the convection movement into small cells for a start.

The Sun's influence is very different on land and sea. Land will warm up and cool down more quickly, while the sea acts as a thermal reservoir to even out heating and cooling. This means that there are stronger convection currents from the continents than the oceans.

Those currents are not just unequal because of their source, but also because the Earth is not a smooth ball. The great mountain ranges will impede or deflect the air currents while the oceans allow much smoother passage.

## Coriolis effect

Next we have what is probably the most interesting physics in the atmosphere: the **Coriolis effect**. This turns air currents (and any other unsteered mass) to the right in the northern hemisphere and to the left in the southern hemisphere.

It's a perplexing phenomenon, which might seem to defy common sense, but applies only to unsteered objects such as winds and ocean currents.

Just standing still on the equator means you are travelling east at about 1,600km/h (1,000mph). Whichever direction you head in, you carry this with you – and it soon becomes faster than the eastward speed of the ground you are now passing over. The effect is an apparent turn away from your initial heading, which continues whenever your latitude changes.

The Coriolis effect is sometimes considered a force, but really is just the manifestation of the spinning of the spherical Earth. It means that winds flowing toward a zone of low atmospheric pressure are turned to the right, and so storms circulate anticlockwise in the north (and opposite in the south).

The Earth's axis is tilted at about 23.4°, and its orbit brings it closest to the Sun on about 4 January each year (so southern hemisphere summers are a little warmer). These factors mean that the planet receives a different amount of solar energy each day.

There are a few other variables to consider when forecasting the weather, including the effect of tides, the varying amount of sunlight reflected into space by clouds, changes to the atmosphere caused by human activity and even the famous "butterfly effect". This suggests a tiny disturbance can propagate into a large unpredicted event.

## Down the drain

What is really interesting about the Coriolis effect is that it acts on any unsteered object, so it also explains the circulation of **ocean currents** and affects the track of missiles and aircraft. If you were flying an aircraft with no autopilot other than a simple device to maintain height with level wings, your track over the ground would turn continuously to the right in the northern hemisphere.

The effect is sometimes thought to make water swirl in opposite directions down plug holes in the two hemispheres. Know-it-alls like to scoff at people making such a claim, but the joke is on them – sort of. It is true that local conditions such as any existing movement in the water and the shape of the bowl will swamp the tiny effect – but there isn't a limit on the size of object where the Coriolis effect can take place.

A series of university experiments in the USA and Australia have demonstrated it in small water tanks a metre or two across, but only after extreme care was taken to ensure uniform geometry and by waiting many hours for all residual currents in the water to dissipate. In these cases, water always drained clockwise in the north and anticlockwise in the south.

On an even smaller scale, many process plants and laboratories have flow meters that make use of the Coriolis effect to provide an accurate measure of the mass flowing in a pipe. These devices are just a few centimetres in size.

## El Niño

Conditions in the Pacific Ocean give rise to a phenomenon that affects the whole globe, on average every five years. Known as El Niño (little boy), it happens when a warmer ocean current develops near the equator in the eastern Pacific. Changes to weather patterns can be serious, with failure of seasonal rains resulting in lost harvests and more frequent tropical cyclones.

Meteorologists use increasingly powerful computer simulations to try to predict what the weather will do next, based on thousands of observations from Earth and from satellites. With so many variables, and given the chaotic nature of the atmosphere, it might not be surprising that it isn't always right.

There is one thing meteorologists routinely get wrong though – the start date of the seasons. The historical scientific basis of this involves the solstices (when the Sun appears highest or lowest in the sky) and equinoxes (when day and night are the same length). These are, within a day or two, 21 December and 21 June, and 21 March and 23 September. Weather forecasters sometimes use an arbitrary start date of the beginning of a month, because they find this more convenient, but that is a poor reason to overturn the solid science of orbital mechanics.

## Measuring the distance to a thunderstorm

Whether you find a **thunderstorm** thrilling or terrifying, you might want to know just how far away it is and whether it is moving toward or away from you. This is as easy as counting.

The flash of a lightning bolt and the boom of thunder are created at the same time in the same place, by the discharge of millions of volts of electrical charge built up as the cloud develops.

The flash travels at the speed of light, which is effectively instantaneous (it actually covers 300km or 186 miles in one thousandth of a second). The thunder moves at the speed of sound, a relatively sluggish 1,235km/h (767mph). All you have to do to measure the distance to the storm is count the seconds between the flash and the boom.

Every three seconds means a distance of one kilometre, and every five seconds means one mile. If you count 20 seconds, that means the storm is about 7km (4 miles) away.

Heat from the Sun sets up convection cells with winds blowing north at high level, south at low level.

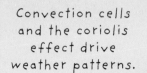

Convection cells and the coriolis effect drive weather patterns.

Coriolis effect turns all unsteered objects in the Northern hemisphere to the right.

Rotation of Earth

All unsteered objects in the Southern hemisphere are turned to the left.

Deflection of winds causes weather systems to rotate in opposite directions in each hemisphere.

# CLIMATE CHANGE

Over the last couple of centuries humans have dug up and burned trillions of tons of fossil fuels, mainly coal and oil, converting enormous quantities of oxygen from the atmosphere into carbon dioxide ($CO_2$) and other gases. This has enabled us to create the modern world in which more people than ever enjoy long lives, good food and improving health care. It has also enabled mechanized warfare on unimaginable scales and caused significant damage to the planet itself.

One of the most pressing issues we face as a species is climate change. All those gases released into the atmosphere are changing the way it behaves, allowing it to trap more energy from the Sun and consequently warming the Earth.

The science behind these effects is solid. The atmosphere is transparent at certain wavelengths (see page 84) and opaque at others – in other words it absorbs some energy from the Sun while allowing other energy to pass. The main **global warming** gas, $CO_2$, is good at absorbing infrared wavelengths – heat energy. It doesn't care whether this comes from above (the Sun) or below (human activity), it absorbs it all the same.

There is surprisingly little $CO_2$ in the atmosphere. Nitrogen and oxygen together make up about 99 percent of our air, and argon nearly all the rest. $CO_2$ is fourth on the list and accounts for only about 0.04 percent – or around 400 parts per million (ppm).

It's a small proportion that has a big effect, in the same way that a dash of chilli sauce changes the taste of a meal. A century ago the $CO_2$ content was about 300ppm, and a century before that about 280ppm. The increase is directly related to the expansion of industry, and it is changing our climate.

One of the most notable effects of this is on **sea levels**, which are inexorably rising. A warmer atmosphere means that the great ice caps on Greenland and Antarctica are shrinking, as are the vast majority of glaciers from Alaska to the Andes and Himalayas. All that water ends up in the oceans, where thermal expansion caused by warming means it occupies a larger volume.

## Sink or swim

Water is unusual in many ways, including not being most dense when solid – if it was, ice would sink rather than float. Water is most dense at about 4°C (39°F), but the average global ocean temperature is now about 17°C (63°F) and climbing. This warmer water takes up more space than cooler water.

The combination of melting ice and expanding water means that sea levels around the world are rising, and the rate of rise is increasing. In the last 100 years, global mean sea level has risen by 10–20cm (4–8 inches) – an average of about 1.5mm (0.06 inches) per year. But in the last 20 years, and with the benefit of better monitoring, the rate has been about 3.2mm (0.13 inches) per year. If the rise steadies at this rate, sea levels will be about 300mm (12 inches) higher at the end of the century than the beginning, with serious implications for many low-lying places, from Pacific and Indian Ocean islands, to major ports including New York and London.

## Fig. 1 – Sea level change 1993–2017

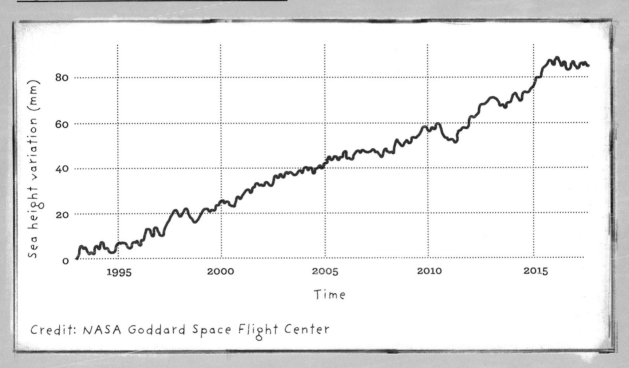

Credit: NASA Goddard Space Flight Center

## Cream or soda?

There are some quite unexpected things that you can do to help the environment. If you have a fondness for whipped cream from a can, for instance, simply giving this up can make a surprisingly big difference.

The propellant in these cans is nitrous oxide, $N_2O$, also known as laughing gas. Its global warming potential is no laughing matter though, being about 280 times as powerful a greenhouse gas as $CO_2$.

Your 200ml (6.8oz) can of cream will do as much damage to the atmosphere as the $CO_2$ from more than 24 2-litre (70oz) bottles of fizzy soda. Whip your own cream, and save the planet.

# RENEWABLE ENERGY

Most of the energy we can use on Earth is ultimately sourced from the Sun, and in human terms the Sun is practically inexhaustible. That doesn't mean that all energy is renewable, because much of what we use today has been locked up for millions of years in the form of fossil fuels – coal, oil and gas.

**B**urning these has all kinds of implications for the future of our planet, but there are plenty of ways we can use the abundant energy delivered to us free each day. Our basic biological energy requirements are met through eating plants (fueled by photosynthesis, a kind of solar power) or animals (fueled by eating plants).

## Turning photons into electrons

That doesn't really help us power a computer, factory or car, but harnessing the power of sunlight can. Photovoltaic solar panels exploit the odd physics of semiconductor materials such as silicon to produce electricity from the photons powering into them at the speed of light. This is the photoelectric effect, and is also seen in the sensor of a digital camera (see page 23).

The secret is a silicon sandwich, generally one layer with added phosphorus, and another with added boron. Phosphorus atoms have a "spare" electron in their outer shell, while boron atoms are an electron short. Where they meet is called a P–N (positive–negative) junction.

When photons from the Sun smash through this sandwich, they liberate spare electrons in the N (phosphorus) layer. These try to make their way to the electron-hungry P layer, but are blocked by the semiconductor properties of the silicon. All that is needed is an electrical connection to complete the circuit.

Hey presto – useful electricity.

There are other ways to harvest power from sunlight, including thermal solar panels, which absorb heat for hot water, and solar concentrators which use arrays of mirrors to focus the Sun's energy. Recent advances in photovoltaic manufacturing have made panels cheaper and more efficient, and as a result they are increasingly common.

## Catch the wind

Solar power is also the driving force behind wind turbines, although a little less directly. Wind is the result of heat from the Sun acting unevenly on the Earth (see Weather forecasts, page 134), and causing great masses of the atmosphere to move over the surface.

Any moving mass contains kinetic energy, and that presents an opportunity to generate power. The physics behind wind turbines is the same as that behind flight, exploiting **Bernoulli's principle**.

This tells us that when a fluid (such as air or water) speeds up, its pressure decreases. The effect allows aircraft weighing hundreds of tons to generate enough lift to leave the ground, simply by traveling fast enough to have the air flowing over its wings at low pressure above and high pressure below.

In the case of a Boeing 747, a mass of just over 400 tons is supported by the atmospheric pressure difference on 540m² of wing (900,000lbs on 5800ft²). Coincidentally, this is a wing loading of 747kg/m², which is about one-fourteenth of the pressure of the atmosphere at sea level. In imperial units this is just about one pound per square inch (lb/in²).

Whichever measurement system you use, this might not seem like a lot to keep a huge machine in the air.

The same effect is what allows wind turbines to extract energy from atmospheric movements. As the wind blows across the device, it travels faster over the more curved blade face and slower over the less curved, setting up a pressure differential that pushes on the blade.

Constrained by the central rotor, the blades turn a generator. This uses electromagnetic induction (see The electric guitar, page 42) to convert the movement into electricity.

In other words, the radiation energy from the Sun becomes thermal energy on the Earth, transferring to kinetic energy in the wind and mechanical energy in the machinery, before becoming electrical energy in the wires and ultimately (for example, when a light is switched on) radiation energy again.

## Fig. 1 – Photovoltaic cell

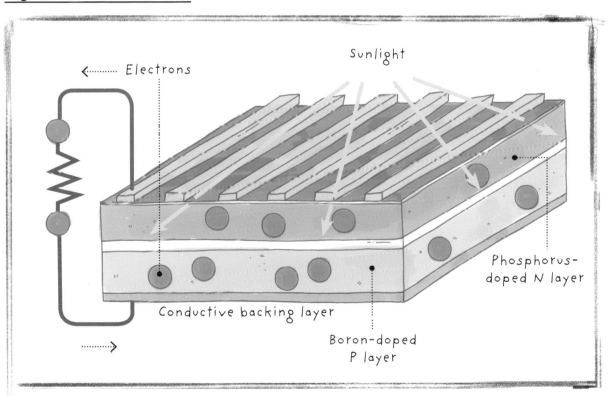

Sunlight

Electrons

Conductive backing layer

Boron-doped P layer

Phosphorus-doped N layer

# Time and tide

The moving fluid masses in the atmosphere are mirrored by the moving fluid masses in the oceans, which present their own potential for renewable energy. Water is about eight hundred times denser than air, and so the same volume traveling at the same speed carries eight hundred times as much kinetic energy.

On the other hand, ocean currents are much slower than atmospheric winds, but not by as much as you might think. The Gulf Stream, which begins in the Caribbean and traverses the Atlantic to northern Europe, has an average speed of around 6km/h (4mph), which is a brisk walking pace. The average wind speed over the UK is around 16km/h (10mph), which is only a little faster than over the USA (14km/h, 9mph).

Where the difference is really felt is in maximum speeds – winds over 160km/h (100mph) are regularly recorded, while the fastest ocean current is just 8km/h (5mph), although exceptional tidal races have been measured at over 30km/h (20mph).

Although not as fast, ocean currents are more reliable than winds and so present an appealing target for renewable energy installations. Even if the current is just 4km/h (2.4mph), a quarter of the average wind speed, the greater mass of water means that two hundred times as much energy could be extracted per square metre of turbine area. The difficulties are in engineering, as all this has to be underwater, and within environmental protection. It would do the renewable energy sector no favours if fish and other creatures were to be injured or killed by submerged turbines.

**Harnessing tides** – powered by the Moon rather than the Sun – may avoid some of these problems. The method is suited to a few specific coastlines with a high tidal range, such as the Bay of Fundy in eastern North America, the Severn estuary in the UK, the Rance River in France and Sihwa Lake in South Korea.

Here the gravitational pull of the Moon raises the sea level by several metres, every 12½ hours. At Fundy the range is up to 16m (52.5ft), and at Severn as much as 15m (49.2ft), and every cubic metre of seawater represents 10kJ of potential energy for every metre it is raised.

As the Earth spins, and the Earth and Moon orbit their mutual barycentre (see Life on Earth, page 90), the oceans are dragged around the surface by gravitational forces.

No energy comes for free, and extracting some from a tidal power station removes it from the Earth/Moon system. Although immeasurably small, the effect of tidal power is to increase the dragging force of the ocean, slowing the Earth's rotation and letting the Moon slip into a slightly higher orbit. Both of these things are happening anyway, but would be (very slightly) increased by a new tidal barrage.

# Where's the wind gone?

Winds around the world are steadily getting lighter – not by much, but by a measurable amount. Cesar Azorin-Molina, a climatologist at the University of Gothenburg in Sweden, says the phenomenon of "stilling" has decreased the average wind speed by about 0.5km/h (0.3mph) every decade since the 1960s.

The reasons are not clear, but may be to do with the "roughening" of the surface caused by the growth of cities, or as an unexpected effect of climate change. One suggestion is that increased cloud cover is responsible.

This might mean lower winds and cloudier days in future, both of which could be problems for renewable energy.

## Fig. 1 – Wind turbines: visible and audible, but free clean power

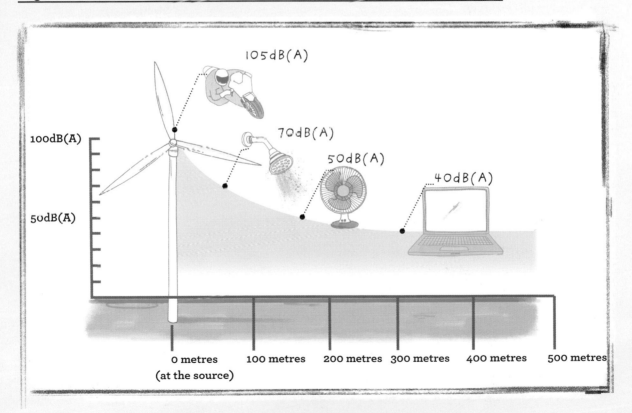

# MIGRATION

Migration is one of the great spectacles of the natural world, with creatures as diverse as butterflies and great whales undertaking annual journeys of thousands of kilometres to find optimum places to feed and breed. Quite how they know the correct direction to head involves a variety of tricks, including a good visual memory, internal magnetic compasses and even quantum mechanics.

**M**any migrations are essentially north–south, between winter and summer habitats. This happens to be convenient for **navigation** as well, as shadows on the ground align in that direction (particularly in the middle of the day). The Earth's magnetic field is also roughly north–south, and many animals are able to use the field for direction finding.

Some of the techniques remain a bit mysterious, and some are quite surprising. Even bacteria – the smallest kind of organism – can include nanometre-sized particles of magnetite within themselves, which orient just like a compass. Many birds, including pigeons, have similar structures in their eyes or brains, but quite how they interpret the information is uncertain.

The European robin (and probably other species) has another trick. Inside its eyes are proteins called cryptochromes, which have been found to include pairs of entangled electrons. Entanglement is a quantum physics effect that means any change in the state of one of the pair is instantly applied to the other, even though there is no apparent connection between them. Entanglement is sometimes called "spooky action at a distance".

The robin's cryptochrome is sensitive to blue light, which provides the power source for its quantum navigation instrument. When the animal turns its head inside a magnetic field (like the Earth's), the cryptochrome shows a chemical response which slightly changes its photosensitivity. This might be seen by the bird as a brighter area of its field of view, indicating the direction to fly.

Human navigators took centuries to work out a good method of east–west navigation, relying on precise clocks that maintained accuracy while on board ship. Another small bird, the reed warbler, has worked out a different method. Instead of simply detecting the direction of magnetic north, it measures the magnetic declination – the difference between magnetic and true north. Warblers that are traveling east–west from Russia to Scotland use the difference between the two norths to provide a kind of avian satellite navigation system to help them on their journey.

## Fig 1 – Getting a lift from the leader

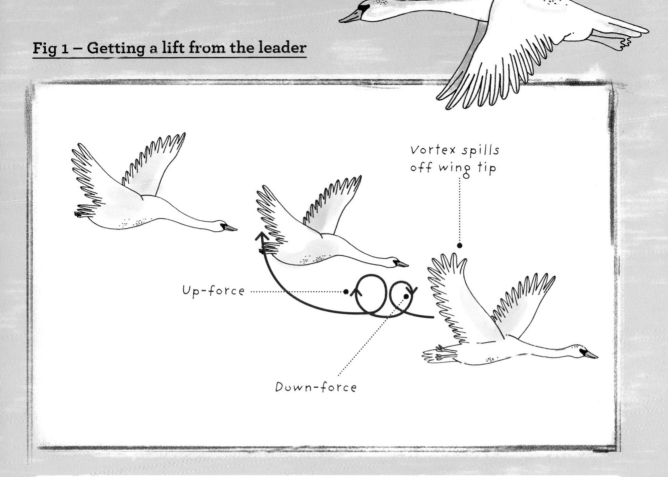

Vortex spills off wing tip

Up-force

Down-force

# Flying "V"

Many long-distance flyers adopt a distinctive "V" formation in flight. This is a clever way to minimize the energy they expend, and also helps to maintain social structures of a flock so that, when they arrive, they are with their tribe.

The lift generated by a flying wing is due to the pressure difference between bottom and top surfaces. At the wing tip, the higher pressure from below spills out to create a **spiral vortex**. This becomes a wake behind the bird, and is a source of considerable drag. Aircraft designers create many different wing tip designs to reduce these tip vortexes, but birds have a way of turning them to their advantage.

As the vortex spirals, it causes a down-force on the inside and an up-force outside. The next bird in the formation rides just behind, to one side, and slightly above the bird in front. This way the tip vortex gives it a little extra lift, reducing its energy consumption.

It might seem hard on the bird at the point of the "V", but they also get a little push forward by the pressure wave of the animals behind – and in any case, they take turns to be leader.

# PEACOCK FEATHERS

Many courtship rituals in the natural world involve spectacular displays, few of them more extraordinary than the peacock. The male bird unfurls a huge fan of brightly coloured feathers, with dozens of mesmerizing eye spots, and shakes his stuff in the hope of attracting the attention of the rather dull-looking peahen.

**W**hat neither of them realize, being bird-brains, is the surprising physics behind his efforts to find a mate.

First, there is the colour. Feathers are one of nature's miracles, highly adapted through evolution to provide a tough protective layer. They are lightweight, often waterproof, highly insulating and partly self-repairing. In many birds they also have a display function. The beautiful blues and greens of a peacock are not the result of brightly coloured pigments, but of **nanoscale structures** built into the feather surface.

The feather is built of a central quill or rachis, from which branch two rows of thin barbs. These in turn sprout two rows of thinner barbules, equipped with tiny hooks which snag other barbules to make the whole feather strong.

## No pigments needed

The barbules have another function, of providing the colour. They are made of keratin, the same substance that makes up human fingernails, with an outer coating of melanin. This is a natural **pigment**, also present in human skin, but in these feathers it creates colour in a very different way.

Pigments usually impart colour by absorbing some wavelengths of light, and reflecting others. It is the *reflected* light that we see. In the case of feathers, the melanin is arranged in arrays of microscopic rods. The precise spacing of these rods determines how light is *refracted* off the feather, and what colour it appears to be. The barbule may appear blue, or green or yellow, just by a difference of a few nanometres in the spacing of these microscopic rods.

# The birds and the beetles

This structural colour is also seen in beetles and butterflies, and is the same effect found when an optical disc (CD or DVD) is held to the light. In this case it is the tiny pits that hold the information (see Optical discs, page 34) which act as a diffraction grating to provide a rainbow of colour.

But while a DVD can display brilliant reds as well as blues and greens, this doesn't happen in nature. When rods are spaced at about 470nm, the feather appears blue, 540nm gives green, while 580nm gives yellow. A red colour would be expected if the rods were spaced at around 700nm, but this longer spacing allows too much interference from shorter wavelengths and the colours end up looking muddy. Birds such as the red cardinal use pigments, not microscopic rods, to generate their bright red colour.

Peacocks have one further trick up their tail fan. Their unusual feather structure produces distinctive eye spots, surrounded by relatively sparsely feathered areas. The distribution of the spots is remarkably clever, as they are positioned at the natural harmonic nodes of the feathers. All objects have a natural frequency, whether they are guitar strings or suspension bridges, and when set into motion they have points that move a lot and points that stay almost still.

A peacock's eye spots are positioned at the nodes, the points that remain still, while the fluffy background feathers shimmer and vibrate when the bird shakes his stuff. The result is quite mesmerizing to the female bird, enhancing the peacock's chances of being chosen as a suitable mate.

Feather barbs interlock with tiny barbules

Barbule surface composed of nanometer-sized rods

# SUNBURN

What we think of as sunlight is really complex multi-band electromagnetic radiation, filtered by the atmosphere. We feel the warmth of the Sun through infrared radiation, and see the landscape through visible wavelengths.

If we stay out too long, especially at higher altitudes or in lower latitudes, we may also see and feel the effects of sunburn. This is a type of radiation burn caused by ultraviolet radiation.

**Ultraviolet (UV)** is just beyond the violet (purple) end of the visible spectrum, with wavelengths shorter than visible light but longer than ionizing radiation including X-rays and gamma rays.

Ionizing radiation has enough energy to knock electrons off atoms, converting them into charged ions. While UV doesn't quite hold this power, it is still energetic enough to cause damage to skin cells, DNA, and eyes.

For convenience, ultraviolet is categorized in three bands, a bit like the colours in visible light. The longest and least damaging wavelengths, between 320 and 400nm, are **UVA** or near ultraviolet. The more harmful **UVB** are 290 to 320nm, while the most powerful 100 to 290nm wavelengths are UVC or far ultraviolet.

Thankfully, the atmosphere absorbs UVC pretty efficiently, so this is of little concern for human health. This is not the case for UVA and UVB, which have both helpful and harmful properties.

First the good news: UV helps the body generate vitamin D, which in turn helps the digestive system absorb calcium. Too little vitamin D can affect the growth of bones and teeth, causing rickets or osteomalacia. So a certain amount of exposure to sunlight is a good thing.

## Varieties of UV

Less helpfully, UV can cause serious damage. UVB radiation carries more energy, and is responsible for sunburn and reddening of the upper layers of the skin.

Surprisingly, given that it has lower energy, UVA wavelengths penetrate deeper into the skin. UVB is stopped by the upper layer, the epidermis, but UVA penetrates through the dermis and can reach the tissue below. About 95 percent of UV light is in the UVA band.

Once in the skin, UV damages cells directly through radiation and by physically cutting strands of DNA. These may be shed or repaired, but some may proliferate and mutate, presenting a risk of cancer.

Besides staying out of the Sun, the best protection is a good sunblock. Health authorities recommend sun creams of at least factor 15, but it may be more important to choose a product effective against both UVA and UVB – many are not formulated for UVA.

The lamps in sunbeds generate proportionally more UVA than in natural light. This may cause quicker tanning but also causes the skin to age and go wrinkly more quickly.

One other underappreciated risk is UV light penetrating into cars. Windscreens are formulated to provide good UV protection, but side windows may not be. This could explain the reported rise in facial skin cancers and cataracts on the left side of people in the USA.

## We saved the planet

The more damaging shorter UVC is absorbed in the stratosphere by ozone, and had the world not taken action to protect this vital shield we might now be dealing with another climate catastrophe. Ozone is destroyed by halogenated chemicals such as chlorofluorocarbons (CFCs) used in industrial processes, aerosols and refrigeration.

When the dangers of CFCs were recognized in the 1980s, the Montreal Protocol called for the gases to be phased out. However, the damage has not been fully repaired. It is thought that ozone depletion will peak around 2020, causing an increase of 10 percent in UV radiation reaching the ground, before falling back to "normal" by about 2050.

## Fig. 1 – Trust me on the sunscreen

UVB rays are associated with sunburn

UVA rays are associated with skin ageing

# RAINBOWS

The legend of the pot of gold at the end of the rainbow has enthralled most children at some time, and is as much a part of the mythology of growing up as the Tooth Fairy and the Easter Bunny. A lesson most of us learn fairly quickly is that you can never reach the end of the rainbow, to claim the gold, because the rainbow "isn't really there".

**W**ell, here's some news. The rainbow is there, and is just as solid and tangible as the raindrops that make it. When we see a rainbow, what we are really seeing is millions of individual droplets falling through the atmosphere, lit by the Sun, and coloured by the refractive properties of water.

You need certain conditions for a rainbow to appear, which means they are much more likely in some places (such as the west of Ireland) than others (the Sahara desert).

First, there has to be enough moisture in the air. A good rain shower should do it. Second, you need a strong source of directional light (usually the Sun, although a moonbow is a rare and beautiful thing). It doesn't take an Einstein to realize that clear sunshine and rain are not necessarily found together very often.

The third and most critical thing is that the Sun and the rainy sky have to be arranged just right, and so do you. If the Sun is high in the sky, the rainbow will appear low to the ground, and if the Sun is higher than 42 degrees you are unlikely to see one at all.

The 42-degree angle is critical, and is the result of a combination of **refraction** and **reflection** of sunlight within the spherical rain drops. A ray of sunlight entering a drop is bent, or refracted, because water is denser than air. As the sunlight bends, it is split into its constituent colours (the spectrum, or "colours of the rainbow") just as if it was being split by a prism or diffraction grating.

The ray is now spreading out like a multicoloured fan within the drop, but soon reaches the other side. Rather than passing through the water/air boundary again, it is now reflected internally before reaching the other side of the drop and undergoing a second refraction.

This sends the light back in the general direction it came from, but having been turned through a total angle of between 318 and 320 degrees. The two-degree difference is the apparent angle between the red wavelengths, refracted the least, and the blue/violet wavelengths which refract the most.

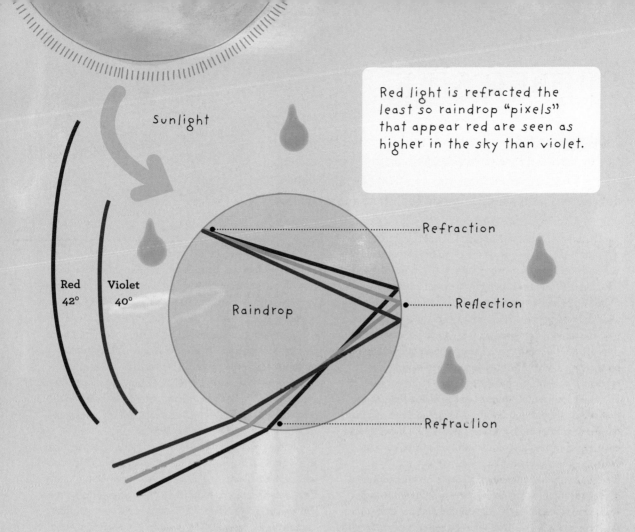

Sunlight

Red light is refracted the least so raindrop "pixels" that appear red are seen as higher in the sky than violet.

Refraction

Red 42°  Violet 40°

Raindrop

Reflection

Refraction

# Every rainbow is centred on you

The 42-degree rule is why rainbows appear as arcs. The shadow of your head is at the centre of a circle, with the red band of the rainbow 42 degrees away from the shadow line. If the Earth didn't get in the way, you would see a complete circle (which is also why you sometimes see a rainbow halo or "glory" around the shadow of an aircraft).

Because every rainbow is centred on the shadow of the observer's head, every rainbow is unique to the observer. You may be standing right next to somebody, admiring the spectacle, but the rainbow they see is not the same as the one you see.

This also explains the myth of the pot of gold. The rainbow is really there, but it stays centred on your shadow and moves when you move. If you race to the place where the rainbow ends, the rainbow races with you. The end is impossible to reach, so who's to say there isn't a pot of gold there?

# SPIDER SILK

Modern science has created an impressive range
of new materials with some extraordinary properties,
but the natural world has its own wonder materials
as impressive as anything cooked up in a test tube.

Spiders produce threads of silk to build their webs and catch prey, creating natural polymers stronger than steel. Quite how they do this has been the subject of research for decades, and is now understood to owe a bit to biology, a bit to chemistry and a lot to physics.

The biology bit involves the glands that secrete the liquid material for the silk. These glands, or spinnerets, have cells which extract protein molecules from the spider's bloodstream and collect them ready for use.

The protein molecules themselves are the chemical part of the production. Proteins are composed of chains of amino acids, which are themselves complex molecules assembled from compounds built mainly of carbon, hydrogen, nitrogen and oxygen.

The proteins in spider silk are some of the longest found anywhere in nature, but are assembled mostly from the two smallest amino acids – alanine and glycine. Alanine forms crystalline sheets which are aligned with the axis of the silk fibre, packing tightly together to provide structural strength.

The amazing elasticity of spider silk is due to the glycine molecules, arranged in repeated sequences of five. These polymers join end-to-end, with a 180 degree turn each time, resulting in a long helix. This looks and behaves exactly like a coil spring.

Spiders produce several different types of silk with different properties, by changing the proportions of amino acids in the protein strands. The most elastic type is known as capture silk, used to trap insects flying into webs. This has over 40 repeats of the 5-molecule glycine unit, and as a result can stretch up to 4 times its original length before breaking.

The type most studied by scientists, because it has the highest strength, is called dragline silk. This is used for the main frame of a web, and for the spider's safety line to escape predators. Here the glycine units repeat only eight or nine times, creating a relatively stiffer and stronger filament. Even so, it can stretch by about 30 percent, which is much more than steel or any synthetic fibre.

## Strong or tough?

Just how strong is it? While it is commonly said that spider silk is stronger than steel (I said it myself, just a few paragraphs ago), this is only partly true.

The **tensile strength** of a material is the load that can be resisted per unit area of a strand before it snaps. A typical high-performance steel has an ultimate tensile strength of about 2,000MPa (megapascals, equal to one newton/mm²). Dragline spider silk has an ultimate tensile strength of... also about 2,000MPa.

There are a couple of other qualities that need to be considered, though. The first is weight, or density. Steel is about six times as dense as spider silk. A 1mm² wire of steel and a 1mm² strand of dragline silk would both snap at a load of 2,000N (call it 200kg or 441lb), but the steel would weigh about 8g (0.3oz) per metre and the silk only 1.3g (0.05oz). A dragline strand of the same weight as the steel wire would hold about 1,200kg (2,645lb) before snapping.

There are synthetic materials with higher tensile strength than either, such as aramid (Kevlar®) at 3,000MPa, and carbon fibre at 4,000MPa. But these super-strong materials have a weakness in their lack of toughness.

The **toughness** of a material is the amount of energy it can absorb before failure, and here the elasticity of spider silk makes it a clear winner. Steel has a toughness of 4,000J/kg, and synthetic fibres about 40,000 – but dragline silk is 140,000!

## Fig. 1 – Molecules in spider silk that make it stretchy and strong

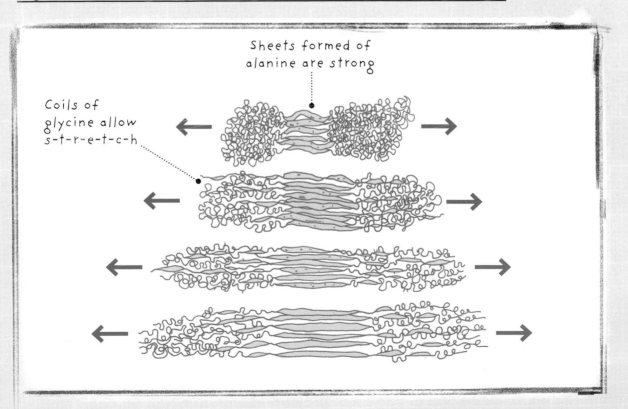

Sheets formed of alanine are strong

Coils of glycine allow s-t-r-e-t-c-h

TRANSPORTATION

# AUTONOMOUS AUTOS

What will be the biggest technological change of the first half of the 21st century? There's a good chance it will be driverless vehicles, which promise a revolution in transportation. Autonomous cars, trucks and taxis will – if the hype is to be believed – make travel safer, quicker, cheaper and better for the environment.

**M**uch of the cleverness of these wheeled robots is contained in the software that enables them to determine where they are, how to get where they are going and how to negotiate obstacles on the way. Critical to this is what we humans consider the moral and ethical decisions that have to be made when things don't go according to plan. Faced with a situation where a crash is inevitable, should the electronics try to protect pedestrians, the occupants of other vehicles or its own occupants?

That's a big programming conundrum, but it can only be tackled if the physics of the situation are properly understood. The computer's situational awareness depends on an array of sensors to recognize and measure the world around it. Each type has its own advantages and disadvantages, and truly self-driving vehicles will have to use more than one type.

**Radar** sensors are quite cheap and well understood. A transmitter beacon emits a microwave signal, typically about 3,900μm (77GHz), and registers any reflected waves it receives. Radar can see through weather, and is reflected strongly by metals. It also reflects off people, but not as strongly, giving a partly translucent image, while wood and painted plastic might not be detected at all. Radars use microwave energy, so any system in a vehicle must have fairly low power to avoid cooking pedestrians and animals. This restricts its range to about 250m (820ft). Radar is the primary sensor used by Tesla cars.

## Seeing in the dark

The main alternative is **lidar** (light radar), which uses laser energy rather than radio frequencies. Lidar continually scans the area around the vehicle with sweeps of laser light, using the time taken for a return signal to calculate distance. It can see up to about 200m (660ft), which is not very far at motorway speeds. Laser wavelengths outside of the visible range have to be selected, to avoid distracting oncoming traffic,

making infrared the lasers of choice. But infrared isn't very good at penetrating snow, fog or rain and this limits its usefulness. Lidar is favoured by Waymo self-driving cars.

**Ultrasonic** sensors are better than radar or lidar for very short distance detection, and are already installed as parking sensors on many modern cars. These work just like a bat's echolocation method, emitting ultrasonic pulses and listening for the echoes. Because these travel at the speed of sound, rather than light, short distances are easier to measure. They're not much use for longer distances and become less reliable at speed.

All of these are active sensors, which emit energy of some sort and make measurements based on what is reflected. Human drivers don't do this, but instead use their inbuilt passive sensors (called eyes). Camera recognition systems mimic this in a technique known as machine vision. This has a longer range than any other method, in good light at least, and is the only

approach that can recognize colour – vital for checking traffic lights. Optical sensors can be trained for character recognition, so they might be able to read road signs.

In parallel with these, an autonomous vehicle will be reliant on satellite navigation and an array of accelerometers and other sensors to provide a constant flow of information.

Despite all these challenges, it is quite remarkable that self-driving vehicles perform as well as they do. The technology isn't completely solved yet, and big questions remain. How will they cope with the realities of a world where road markings and surfaces are not always well maintained, and where meatbag drivers do unpredictable things?

It is also inescapable that, the more complex a system is, the more ways there are for it to fail. But the driverless car and truck do seem to be coming to a road near you, soon.

Radar emits waves which are reflected back from objects up to 250m (820ft) away.

Self-driving vehicles will rely on a variety of sensors, including at least radar and/or lidar plus satellite navigation, ultrasonics and computer vision.

Ultrasonic sensors emit pulses and listen for the echoes. These can detect objects up to 5m (16 feet) away and are useful at low speed.

# HYPERLOOP

Hyperloop is an attempt to overcome some of the basic problems of physics that make high-speed transport so difficult and expensive. It is also an illustration of how the solution to one problem can create new problems.

The main obstacle to moving fast in the Earth's atmosphere is drag, caused by air resistance. Drag has little effect at low speeds, but becomes progressively worse at higher speeds. Double your speed, and drag increases four times. Double it again, and it's now 16 times the original. It soon becomes the only thing slowing you down.

Drag is caused by all the air molecules that need to be pushed out of the way, and one answer to the problem is to build a streamlined shape that slips more easily through the air. An alternative is to have less air to push away. Airliners fly at high altitude and have very streamlined shapes for exactly these reasons.

But it isn't always convenient to climb to 10,700m (35,000ft), especially for short journeys. Hyperloop promises to solve this by creating a very thin atmosphere at ground level. Passengers will travel in pressurized pods that will hurtle along tubes pumped free of almost all air.

The proposal is for the pressure inside the tubes to be just 100Pa, equivalent to the atmosphere at 46,000m (150,000ft), almost half way to the official edge of space. This would mean just one-thousandth of the sea level atmosphere would have to be pushed aside by the capsules.

Problem solved? Not quite. To achieve the high speeds promised (about 1,000km/h, 600-700mph) means overcoming another barrier of physics, the **Kantrowitz limit**. As the capsule is propelled along the tube, the thin atmosphere in front becomes compressed. As it approaches the speed of sound, it begins to "choke" the tube and effectively put the brakes on.

## Canned heat

In the open atmosphere, an aircraft could accelerate through this pressure wave, creating a sonic boom. Inside the Hyperloop tube, that would make things very complicated. The proposed answer is to install a powerful turbine at the front of the pod, to pump the air to the back. Some designs suggest using some of this compressed air to make the capsule float inside the tube, rather like an air hockey table in reverse. Other designs concentrate on magnetic levitation (see page 160) along with linear motors to raise and propel the capsule. In both cases, the clearance between vehicle and tube wall would be just a millimetre or so.

While the prospect of traveling a few hundred kilometres in only minutes is very appealing, Hyperloop still has many problems to solve. The physics may work, but the real challenges will lie in engineering.

## Fig. 1 – Faster than a speeding train

From Toronto to Montreal
640km (400 miles)

Toronto to Montreal journey times:
Car = 5+ hours
Train = 4+ hours
Plane = 1 hour 10 minutes
Hyperloop = 39 minutes

## Life's a drag

The formula for drag looks quite complicated, but tells us some important facts.

$$\text{Drag} = \tfrac{1}{2}\,\rho\,v^2\,C_d\,A$$

Starting from the right, A is simply the area of the object, and so is a constant. $C_d$ is the coefficient of drag of the object – a measure of how streamlined it is – and is also a constant.

Now from the left: ½ is just a number, and ρ (the Greek letter rho) is a measure of how dense the fluid is. In the case of Hyperloop, the fluid is air and the density is one-thousandth of the density at sea level. Immediately the advantage of pumping most of the air away can be seen.

The final element is $v^2$, which is the velocity squared. This tells us that double the speed means four times the drag, and so on.

The central part of this equation, $\tfrac{1}{2}\,\rho\,v^2$, occurs in many important formulae in fluid dynamics, engineering, aviation and even plumbing.

To get past the Kantrowitz limit, a powerful fan blows air from the front to the back of the capsule, diverting some of it down to the "air hockey" suspension system.

Some Hyperloop designs include magnetic levitation and propulsion.

## Is it new?

Hyperloop is associated with entrepreneur and technology enthusiast Elon Musk, whose business ventures include Tesla electric cars and SpaceX rockets. But the idea of a "vactrain" or atmospheric railway has been suggested many times in the past. The earliest was probably George Medhurst in 1799!

# MAGLEV

If the invention of the wheel was the first world-changing transport technology, the invention of a train that needs no wheels might be the next.

**M**agnetic levitation (maglev) trains use magnetic repulsion so their carriages float a few centimetres above a track, making no actual contact and therefore promising friction-free locomotion. The physics of maglev is superficially straightforward, but quickly diverts into some fascinating areas of materials science including superconductivity and linear motors. There are really two problems that need to be solved: to make the train levitate above the track (while being stable so it doesn't come off at the first bend), and to propel the floating train at a useful speed. And then stop safely at the destination, of course.

Basic magnetic levitation is fairly easy to arrange, making use of the repulsion between similar magnetic poles (north to north, south to south). You might have seen this effect in novelty products such as floating lightbulbs, but this approach is unstable. At scale, some careful engineering with levitation magnets acting vertically and guidance magnets acting horizontally can solve this issue.

An alternative to this **electromagnetic levitation** is electrodynamic suspension, which exploits some cutting-edge physics. Some materials (often ceramics) demonstrate superconductivity when cooled to very low temperatures. Electrical current can flow with zero resistance, so electrons will continue to circulate forever – or until the temperature warms above the transition point.

These superconductors also exclude magnetic fields from their interiors, essentially refusing to allow the field lines to penetrate them. This is known as the Meissner effect, and means that the superconductor repels itself from any magnetic field.

There is a further trick, which is even more useful in maglev applications. If the superconductor is cooled below its transition temperature while already in a magnetic field, it tries to lock itself in its position. It does not simply levitate, but resists any up, down, or side-to-side motion. This phenomenon, called flux pinning, means that a simple flat track will provide vertical and lateral support – the train can still hurtle along at high speed, but will be magnetically glued to the bends. Should the situation arise, a flux-pinned carriage would even remain levitated below an upside-down track. The potential for roller coasters is enormous.

## Do the locomotion

Train carriages floating above a track are all well and good, but modern life demands speed. With no wheels to provide propulsion, how is a maglev train moved forward? The answer is the linear induction motor. This is very similar to the standard electric induction motors that drive everything from your washing machine to your (one day soon) electric car. Instead of being arranged around the central spindle, a linear motor

## First maglev

There are maglev services in China, Japan, Korea and Germany, but the world has almost forgotten its first one. Running a distance of just 625m (2,050ft) between Birmingham airport and Birmingham railway station in the UK, it opened in 1984 and closed in 1995. With no moving parts it should have been an ultra-reliable shuttle service for passengers, but in reality, it suffered frequent breakdowns. It was replaced by a conventional light railway, complete with wheels.

has the same components unwrapped and laid out flat. Electrical current moving in a coil generates magnetic force which imparts a horizontal thrust.

In practice, the two operations of levitation and propulsion are incorporated into the same array of electromagnets or superconducting magnets, giving the train the ability to lift and drive away.

This might all seem like a lot of effort to replace well-proven rail technology, and it is true that maglev railways are quite expensive to build, but having no moving parts, they should be cheap to maintain. The frictionless travel also means better efficiency as well as a smoother and quieter ride. What is sometimes overlooked, though, is that rail–wheel friction becomes insignificant at high speed, where aerodynamic drag becomes the dominant problem.

## Maglev vs Hyperloop

New high-speed public transportation networks might use either of these approaches, or conventional high-speed rail. Hyperloop should be the fastest, but will also probably be more expensive to build, and remains unproven. Some Hyperloop designs incorporate magnetic levitation. Maglev systems have been shown to work, but the benefits they offer may not be worth the additional cost. It's possible the race might be won by fast trains like the TGV and Shinkansen (Bullet Train), which don't need to reinvent the wheel.

## Fig. 1 – Maglev train

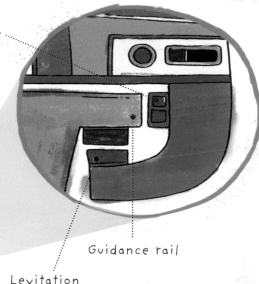

Guidance electromagnet

Guidance rail

Levitation electromagnet

# SATELLITE NAVIGATION

The principles behind satellite navigation might seem straightforward, but there is some pretty far-out physics involved. The routine task of getting a geographic fix from a constellation of orbiting radio beacons involves subtle adjustments for atmospheric effects, special and general relativity, and some sophisticated orbital mechanics. It also requires super-accurate atomic clocks.

The first use of radio for navigation came over a century ago, in an arrangement rather like a lighthouse – a simple position indicator. When radar came along, it added distance measurement to direction to give a fix of position rather than just a bearing (radar stands for radio detection and ranging).

Satellite navigation dispenses with the direction element and instead makes calculations based on distance alone. Knowing that radio waves travel at the speed of light, if you can measure the time taken for a signal to travel from a satellite to a receiver, you can work out the distance. That sounds simple.

The first problem is that the time measurement has to be very precise. Light travels at around 300,000km (186,000 miles) per second, so a regular 1/1000th second stopwatch is of no use. Even if you could measure the time to a millionth of a second, you could only measure the distance to within 300m (984ft). That might be enough for a ship on the open sea, but would be little use navigating city streets.

## Atomic time on the cheap

Atomic clocks register the vibrations of electrons in atoms of caesium-133, and are accurate to about one second in 300 million years. Currently they are also far too expensive to build into a smartphone or satellite navigation receiver. How then does a cheap device check the time signal from a satellite to an accuracy of hundredths of a millionth of a second?

It doesn't. What it really does is compare the time signals from a number of satellites, and performs some arithmetical gymnastics. It's convenient to think that the signal from the first satellite gives a position fix somewhere on a sphere, the second somewhere on a circle where two spheres intersect and the third at one of two points where three spheres intersect. This isn't quite what happens, because one satellite by itself provides practically no position information.

With two signals, time stamped and synchronized with each other, the difference between them can be measured. With three signals a rough fix becomes possible, but it takes a fourth before we get proper navigational usefulness.

The signals are encoded to include the identity of the satellite and its orbital details along with what's called ephemeris data, detailing where the satellite actually is rather than just where it's supposed to be.

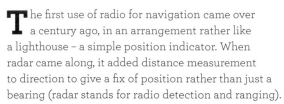

No orbit is completely predictable and even at the relatively high altitude of these devices there is a little atmospheric drag to be compensated for.

Other compensations to be worked in include an allowance for the speed of the satellite, which according to special relativity makes the clocks appear to run a bit slow. Meanwhile the gravity gradient on the radio signal's run down to Earth makes the clocks appear to run fast, according to general relativity, requiring a completely separate adjustment.

Finally, the speed of the signal is not constant, but is slowed down a bit by both the ionosphere and troposphere.

## What's in a name?

The first and best known satellite navigation system is GPS, built by the USA and using a constellation of 24 satellites.

There are at least two other systems in public use, GLONASS (Russia) and Galileo (European Union), with 24 and 30 satellites respectively.

Both China and India are developing their own system, and some receivers gain more accuracy by getting their fix from more than one constellation.

## Fig. 1 – Satellites in Medium-Earth Orbit (MEO)

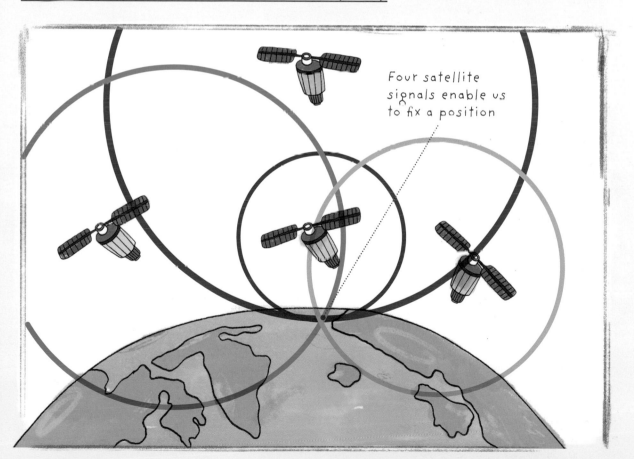

Four satellite signals enable us to fix a position

# MOTOR SPORT

There are two key factors that determine how fast a racing car or motorcycle can make it around a circuit. The first is power, or more specifically the power-to-weight ratio of the machine, which governs how fast it can accelerate. The intimate connection between power, mass and acceleration are revealed by Newton's second law of motion F = ma or force equals mass times acceleration. If you want more acceleration – and racers always do – you need more force or less mass. Preferably both.

Force and power sound like the same thing, but they are quite different. Force is the amount of push applied to an object, at any moment in time, measured in newtons. Work is the force multiplied by the distance (in joules, or newton-metres), and power is the rate at which work is done (watts, or joules per second). 746 watts is one old-fashioned horsepower, still a commonly-quoted unit of power, and Formula 1™ engines under present regulations can generate over 850 horsepower. This is several times the power of a typical family car, propelling a vehicle with a less than half the mass. This is how a Formula 1™ car can accelerate from a standstill to 100km/h (62mph) in a couple of seconds.

It is at about this speed that the second factor, aerodynamics, kicks in. With road-going vehicles, careful contouring aims to reduce drag, so that the power required (and fuel used) to reach and maintain speeds is reduced.

In racing, the primary aim is to keep the car on the track so that it can get around the bends at high speed.

**Downforce**

**Aerodynamics** are all about generating downforce to push the car onto the asphalt, with front and rear wings and a variety of winglets, vanes, fences and flaps. Regulations on what is allowable change regularly, and the teams build cars to different configurations for different circuits, but the overall effect is that F1™ cars have a coefficient of drag ($C_d$) of around 0.9 to 1.1. This is much worse than a pick-up truck (0.6) or family car (0.3) and about the same as a brick.

All that drag creates so much downforce that an F1™ car could drive upside down through a tunnel, carrying ballast equivalent to its own weight again. But it would have to maintain about 160km/h (100mph) to avoid dropping to the ground in an untidy mess.

If you took the wings off and made it more streamlined the car would have a higher top speed, but would not be able to brake so strongly and would spin off on the corners.

## Downforce

## Flying bikes

Teams in the two-wheeled equivalent of Formula 1™, MotoGP™, also experiment with aerodynamic modifications in the quest for an advantage. They face similar problems, plus a few extra ones of their own.

Aerodynamic winglets were controversial when introduced in 2016, partly because of safety concerns in the event of a crash. Winglets and modifications to aerodynamic fairing (the shell over the frame of the bike) help reduce the tendency to wheelie, keeping the bike stable under hard acceleration. But when the bike is leaning over at 60 degrees from vertical, the "down" force is acting more sideways than downwards and not all riders feel there is an advantage.

Modifications to generate downforce increase drag significantly, reducing the top speed of MotoGP™ bikes by 10km/h (6mph).

The $F = ma$ equation is possibly even more important in MotoGP™, because rider weight is not included in the qualifying weight of the machine. Lighter riders gain an acceleration advantage over others on identical bikes, but may lose some of this when it comes to braking.

A modern F1™ is a lot more complex in shape and the downforce is generated in several places. The most important are the front and rear wings.

# GOING *RREEAALLLLYY* FAST

One thing that has marked the development of technology is that we always want to go faster. The wheel was an early invention to help us do this and one that led to carts, bicycles, railways and cars. Before long the motorcycle was invented. When did motorcycle racing begin? At the exact moment the first motorcyclist in history met the second.

T oday many people have cars capable of twice the legal speed limit, but ironically average traffic speeds in cities may be no faster than the horses and carts of the 19th century. Never mind, because we have aircraft that whisk us off on business or holiday at 900km/h (550mph), and countries blessed with high-speed rail have trains regularly hitting 300km/h (186mph) or more.

## The need for speed

None of this is good enough, though, and there are always plenty of proposals for new ways of going farther and faster. The only thing stopping them – besides money – is physics.

The main problem is the atmosphere, which slows down everything trying to pass through it. Atmospheric drag is of little consequence at speeds up to about 80km/h (50mph), but then quickly becomes the main issue designers of high-speed craft have to deal with.

## Drag race

The drag depends on the speed, the density of the air (and so is less at altitude) and the shape of the craft. Slippery streamlined shapes have a low **coefficient of drag**, $C_d$, so where a pick-up truck might have a $C_d$ of 0.6, a typical modern family car might be 0.3 and an airliner only around 0.025.

As you go faster, drag increases. Drag is a function of the square of velocity, so doubling the speed brings a four-times increase in drag. Double it again, and you have 16 times. What's worse is that the power necessary to overcome drag increases with the *cube* of velocity, so doubling speed demands eight times the power, and doubling it again requires sixty-four times. This apparently bizarre situation comes about because the **work** done at double the speed takes place in half the time, and therefore twice the **power** is necessary.

What this comes down to is that the laws of physics conspire against high speed, and demand a high price when we want to have it.

# Boom boom time

As velocities increase further, the atmosphere has another trick up its sleeve. Moving through the air means that the air molecules have to move aside to let you pass. The moving molecules travel in a pressure wave, and pressure waves in the atmosphere are also known as sound waves. They travel at a fixed speed, the speed of sound, which is about 1,235km/h (767mph). The speed of sound is lower at higher altitudes, but this is mainly an effect of temperature rather than pressure. At 0°C (32°F) it is 1,192km/h (741mph). As an object accelerates toward this speed, the air molecules begin to have difficulty moving aside. They pile up at the front of the object, increasing atmospheric pressure, and creating a barrier to further acceleration – often called the sound barrier.

Powering through this barrier requires a lot of effort, and really needs a very slim pointed shape. Aircraft pilots who first approached the speed of sound found that some aircraft became dangerously unstable, while their controls such as ailerons and rudder behaved oddly.

Speeds from around 80 percent of sound, or Mach 0.8, are known as transonic. The air flowing around an object is not all traveling at the same speed, because the shape of the object means air accelerates through certain parts and piles up in others. In the transonic zone, some parts of an aircraft may be supersonic, while other parts are still subsonic.

Once the entire airframe is supersonic, a great deal of stability is restored and pilots have better control. At these speeds each part of the aircraft is aerodynamically independent of the others, because any pressure changes from one part can only travel at the speed of sound and so are left behind before they can affect other components.

The sound wave at the front of the aircraft, meanwhile, has transformed into something rather different – a shock wave. Shock waves travel faster than sound, and take on the shape of a cone spreading backward from the front of the supersonic object. When this shockwave sweeps across the ground, the "sonic boom" is heard.

Civilian supersonic flight was a reality between 1976 and 2003, when Concorde flew regular (but expensive) routes across the Atlantic in less than half the time of conventional airliners. A new generation of supersonic aircraft may take to the skies in the near future, but must find ways to reduce the effect of the sonic boom if they hope to pass over populated areas.

## Fig. 1 – Sonic boom

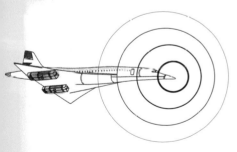

1: Air in front of the aircraft is compressed, moving away in a pressure wave that travels at the speed of sound.

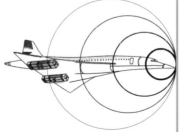

2: At Mach 1, the air pressure builds up because the pressure wave cannot escape.

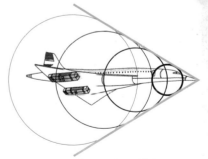

3: Beyond Mach 1 it has become a shock wave trailing in a cone behind the aircraft. The loud sonic boom is heard as the shock wave sweeps over the ground.

# Going hypersonic

Supersonic still isn't fast enough, is it? Two-and-a-bit times the speed of sound, as Concorde managed, means it still takes *hours* to get from New York to London. If we want to do it in minutes, we need to go hypersonic.

At speeds of about Mach 5 and above, the atmosphere begins to demonstrate yet another weird physical effect called dissociation. This is when the brutal impact of a penetrating aircraft is enough to actually break the molecules in the air.

The atmosphere is mainly nitrogen and oxygen, and oxygen is the main problem both because it dissociates earlier, and because the broken molecules of atomic oxygen are very reactive. Friction with the air means that hypersonic objects get very hot, especially on leading edges, and the aluminium alloys used in conventional aircraft become too soft. The additional corrosive effects of atomic oxygen mean that exotic materials such as carbon nanotubes may be needed to maintain structural integrity.

Finding ways to power hypersonic craft is another challenge. Turbofan engines fitted to typical airliners can operate up to about Mach 1.5, and with the addition of an afterburner (as on Concorde and the SR-71 spy plane) can pass Mach 3. A ramjet engine relies on forward motion to compress air for combustion, instead of spinning turbines, and so can generate thrust up to around Mach 6 – but doesn't work at all at low speed and so cannot power a hypersonic aircraft by itself.

Some definitions of hypersonic put the threshold as the speed where a ramjet can produce no thrust, because its exhaust gas is slower than the flight speed. For this, you need a scramjet, or **supersonic combusting ramjet**, in which the air remains at supersonic speeds all the way through the engine.

# Bloodhound

One of the greatest achievements of speed freaks anywhere was breaking the sound barrier at ground level. Thrust SSC did this in 1997 with Andy Green at the controls, setting the world land speed record at 1,228km/h (763mph).

Still, not enough.

The successor car, Bloodhound, aims to be the first to go faster than 1,600km/h (1,000mph). The challenges to overcome are many, and complicated. The shock wave from supersonic travel can penetrate the desert where the run is planned, converting a reasonably solid surface into a fluidized mess that makes control difficult. The wheels will be spinning so fast that they have to withstand a radial acceleration of 50,000g, and aerodynamic effects on the wheels will be much stronger than the interaction with the desert floor when steering is attempted. Even the pilot, Wing Commander Green again, must cope with acceleration and deceleration forces of up to 3g on each run.

The real question is this: will there be any point to further speed records if Bloodhound is successful?

# STEALTH

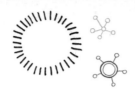

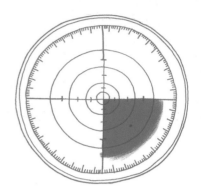

One good way of keeping your military aircraft and ships safe from enemy action is to make them as hard to detect as possible. In modern warfare that generally means making them stealthy, with little or no radar reflection.

A number of different techniques are used to do this, but the two most important are the shape of the vessel and the material it is surfaced with. The intention with both is to prevent a radar signal being reflected toward a receiver.

Radars use the same part of the electromagnetic spectrum as microwave ovens and Wi-Fi®, because these relatively long waves can penetrate the atmosphere better than shorter wavelengths. The penalty is that longer wavelengths (lower frequencies) are less accurate and require larger antenna dishes.

Metals are good radar reflectors, but reflectivity is not the same as **retroreflectivity**. If a radar signal is reflected away from the receiver, it won't be detected. Retroreflectivity turns a signal back the way it came, producing a strong radar return. Retroreflectivity is the trick behind hi-visibility clothing, sending light back toward its source.

The key is to eliminate as many right-angle junctions as possible. Three mirrors arranged so that they form an inside corner of a cube will be highly retroreflective. The Apollo missions left reflectors like this on the Moon for exactly this reason, so they could be detected from Earth.

## Flat shapes

Stealthy aircraft and ships are often noticeable for their slabby appearance, with large flat areas meeting to form convex shapes. This reduces their radar retroreflectivity, sending signals away from detectors and harmlessly up into the sky where they are less likely to be detected. Non-stealthy designs have more complex shapes and inadvertently have many hollows which make them highly visible to radar. Beneath the skin, internal structures are also designed to reduce retroreflective junctions.

Another trick is to keep the "radar-bright" objects such as bombs and radio antennae inside the fuselage as much as possible to hide them from view. One problem here is that opening bomb bay doors can suddenly make a stealthy aircraft much more radar-visible.

Other stealthy design features include having engine intakes and exhausts above the fuselage to reduce the infrared heat signature, and joining panels in a zig-zag pattern rather than straight lines. None of these things are helpful to the performance of an aircraft, so stealthy machines tend to have lower payloads and top speeds, along with higher fuel consumption and hence lower range.

## Paint it black

The second approach is to use a coating to absorb as much radar energy as possible. Just as a green leaf reflects green light but absorbs red, a radar-absorbing coating will absorb radio and microwave energy.

Many of these coatings contain beads or fibres of ferrite, an iron compound. When these are illuminated by a radar signal, they vibrate and covert the energy into heat. This can be quite effective, but the coatings are heavy and this further reduces aircraft performance. New generations of radar-absorbing coating may make use of hollow ceramic beads coated with metals to reduce this problem.

Stealth and detection are at the cutting edge of military science, with each improvement in one usually being followed by a further development in the other. A high-stealth aircraft such as the F-22 fighter is claimed to have a front-on radar cross-section equivalent to a bumblebee.

And no radar operator will think it is suspicious that a bumblebee can fly at more than twice the speed of sound.

Fig 1

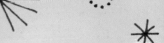

## Fig. 1 – Stealthy by design

Bet you can't see me! There is hardly a right angle to be found in the F-22.

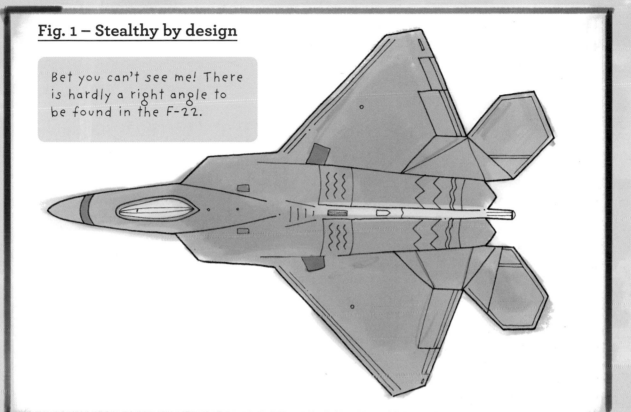

The development of radar in the 1930s and 40s changed the nature of warfare, giving the huge advantage of being able to detect enemies at a distance. Radar works by transmitting radio or microwave energy, and listening for the echo to determine the direction and distance to the reflecting object. The name radar comes from "radio direction and range".

EVERYTHING ELSE

# CURVE BALLS

Both baseball and cricket involve hurling a solid ball at high speed toward a player who thwacks it with all possible might using a wooden or metal bat. In both games the objective is to score more runs than the other team. The real difference between them is the way the bat is held – vertically in cricket, horizontally in baseball – which brings about a fundamental difference in the physics behind them.

It's all to do with the way balls behave as they zoom through the air. No matter how fast the ball is propelled, gravity will pull it into a downward trajectory. At the same time, atmospheric drag puts the brakes on and slows it down.

If the ball is thrown without any spin, that's about it. Throw, thwack and the crowd roars. Add a little spin, though, and things start getting interesting.

Just as the air pulls on the ball, creating drag to slow it, the ball pulls on the air and causes its pressure to change. With a non-rotating ball, a high-pressure zone at the front, and low-pressure zone behind generates the braking force – the drag. When the ball is spinning, the high- and low-pressure zones are pulled into new positions, so that the drag force acts at an angle and changes the ball's flight path.

The part of the ball spinning forward makes the air bunch up and compress, generating high pressure, while the part spinning backward stretches the air creating low pressure. It's called the **Magnus effect**, and allows a skilled arm to make the ball swerve in mid-air.

## Pitcher this

This is where the two sports differ. In baseball, the pitcher is most interested in outfoxing the hitter by delivering the ball higher or lower than expected. A relatively small up or down shift is more likely to result in a missed swing than any left-right movement. Applying topspin to the ball means that high pressure builds up above, low pressure below and the ball arrives a bit lower than the hitter might expect. Backspin does the opposite.

Spin is also important in cricket, but the situation is complicated by fact that the ball bounces before reaching the batter, and by the large raised seam around the circumference of the ball. Confusingly, a bowler using aerodynamic effects in cricket is called a swing bowler, while a spin bowler is doing something quite different by making the ball bounce unhelpfully. A swing bowler will deliver the ball somewhere to the right or left of where the batter expects, ideally causing a missed swing but with the ball on track to hit the stumps.

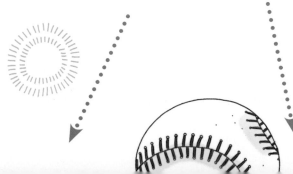

# Throw a frozen rope

A ball with a huge amount of backspin can generate enough Magnus force to counteract gravity, and travel in an almost-straight line (the proverbial "frozen rope"). In baseball and cricket the size and weight of the ball put this beyond human capabilities, but the effect can be achieved in sports such as table tennis and football (soccer), where a relatively light ball can float in a way guaranteed to confuse the opposition.

The ball's surface texture plays a part in this. The dimples on a golf ball might look like they would slow it down, but in fact they allow airflow to remain attached and so reduce turbulence and drag, allowing it to go farther. They also boost the Magnus effect enough that the ball can travel twice as far with dimples as it would without.

Footballers seem to love to complain, and have a new opportunity every four years. Each World Cup competition uses a new official ball, and each time it has different aerodynamic properties than anything seen before. With identical balls for all matches there should be no unfairness, but players do have to re-learn how to make it swerve.

## Fig. 1 – Curve balls

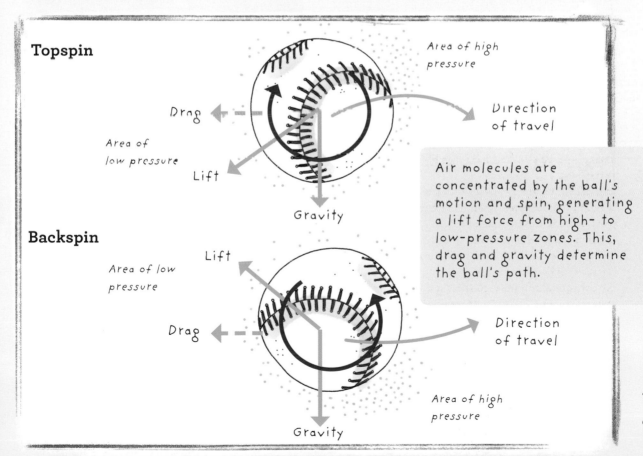

**Topspin**

Drag
Area of low pressure
Lift
Gravity

Area of high pressure
Direction of travel

Air molecules are concentrated by the ball's motion and spin, generating a lift force from high- to low-pressure zones. This, drag and gravity determine the ball's path.

**Backspin**

Lift
Area of low pressure
Drag
Gravity

Direction of travel
Area of high pressure

# THE MPEMBA EFFECT

Scientists love things that defy common sense, because it gives them something to fret about and, perhaps, make some worthwhile discovery. As science fiction writer Isaac Asimov said: "The most exciting phrase in science is not 'Eureka!' but 'That's funny…'."

The **Mpemba effect** certainly falls into this category. It seems so unlikely that it must obviously be false, and yet it can be demonstrated and so must be real. Take two identical containers of water, one hot and one cold, and put them in a freezer. Which will freeze first?

Obviously the cold water will freeze first. The hot water has to cool to the temperature of the cold water before it can continue toward ice, so it must take longer, right? Wrong. Sometimes (but not always) the hot water freezes first. This is the Mpemba effect, sometimes called the **Mpemba paradox**. This is a tortoise-and-hare race that you can try at home, and make your own attempts to explain.

## Whistling in the wind

You sometimes see a TV reporter standing outside in winter, throwing a cup of boiling water in the air to "demonstrate" the Mpemba effect. They are doing nothing of the sort. They are actually showing how a large surface area and very cold air allows water to cool rapidly, turning into a sort of artificial snow.

They also sometimes get it wrong, with painful results. Tip for TV people: Always stand with your back to the wind, no matter what the camera operator tells you. If you don't, you might just throw a cup of boiling water in your face.

## A scientific mystery

Plenty have tried to come up with a convincing explanation for the Mpemba effect, but none have yet succeeded. Theories include the lower density of hot water (meaning that the same volume could have less mass and so freeze more quickly), the suggestion that evaporation reduces the mass of the hot water, or that heating drives off dissolved gases such as carbon dioxide and this changes the freezing point. Some suggest that heated water actually has a higher freezing point than cold water, because the latter will supercool (stay liquid at temperatures below 0°C). It might even be a combination of factors.

Whatever the actual explanation, the Mpemba effect remains a proper scientific mystery.

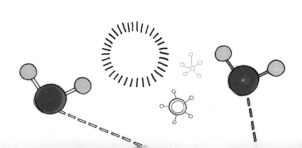

## Who is Mpemba?

The effect is named after Erasto Mpemba, who noticed the odd behaviour in ice cream mixtures while at school in Tanzania in the 1960s. He asked a visiting physics professor, Denis Osborne from University College in Dar es Salaam, to explain it.

Legend has it that young Erasto was ridiculed by classmates for the suggestion, but the professor was interested enough to perform some experiments – and confirmed the effect. The two published a paper on the phenomenon while Mpemba was still a teenager, and science history was made.

If nothing else, this shows that apparently silly questions can still lead to new discoveries.

## Fig. 1 – Water freezing rates

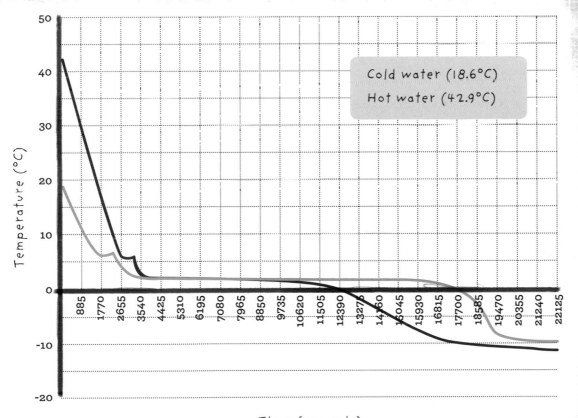

Cold water (18.6°C)
Hot water (42.9°C)

Temperature (°C)

Time (seconds)

# WHY NORTH IS REALLY SOUTH

You might have a good idea of which way is north, and if in doubt you would probably reach for a compass or look at the direction shadows were pointing for clues. But what do we really mean by north?

The simple answer is that north is the direction to the North Pole, which is one end of the axis the Earth spins around. This is **true north**.

But north comes in many flavours, including **grid north** and **magnetic north**. The introduction of digital mapping and navigation means there may also be a separate "map north" for the system you are using (Google Maps use "Google north", for example!).

Grid north may be slightly to the left or right of true north, as a result of the way cartographers have to stretch maps to make a curved surface (the Earth) fit on a flat piece of paper. Take a "square" geographical unit such as the state of Colorado. It's a mapmaker's delight, being defined by lines of latitude at 37°N and 41°N, and longitude from 102°03'W to 109°03'W.

## Making it fit

None of those lines is straight though, because the Earth is (roughly) spherical. To make it fit neatly on a rectangular piece of paper means making a few adjustments. The state's southern border is 623km (387 miles) long, but the northern border only 589km (366 miles) – a difference of 34km (21 miles).

The difference between grid north and true north will be marked on any navigational map, as will the magnetic declination or variation. Compasses point to the magnetic north pole, which is presently located somewhere around Ellesmere Island on the far north of Canada. This pole is continually moving, at a rate of over a kilometre a week, so maps can soon become outdated.

## Fig. 1 – Magnetic variation across the world

| City | Magnetic variation 2017 | Rate of change per year |
|------|------------------------|------------------------|
| **London** | 00°21'W | 10' east |
| **New York** | 12°58'W | 1.5' east |
| **Los Angeles** | 12°2'E | 4.8' west |
| **Hong Kong** | 02°50'W | 3.8' west |
| **Cape Town** | 25°04'W | 4.8' west |

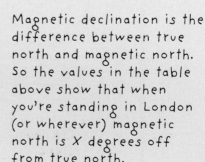

Magnetic declination is the difference between true north and magnetic north. So the values in the table above show that when you're standing in London (or wherever) magnetic north is X degrees off from true north.

Airport runways are numbered for their magnetic bearing, rounded to ten degrees – so an aircraft on runway 09 will take off and land approximately to the east, 090 degrees, while the same stretch of concrete in the opposite direction will be runway 27 (west, or 270 degrees). This makes obvious sense for navigation purposes. But the moving pole means runways have to be redesignated from time to time. London Heathrow used to have main runways labeled 10/28 (one runway running east at 10 degrees and the other running west at 280 degrees), but since 1987 these have been 09/27 due to the magnetic declination. The two numbers are always 18 different, as the directions are 180 degrees apart.

Now, what if I told you that the magnetic north pole was, in fact, nothing of the sort? We know that magnets have a north and south pole, and that similar poles repel each other while opposites attract. Every magnetic compass has a "north" pole, which would never point toward another north pole but always to a south pole. It's obvious, then, that the magnetic north pole is actually the south pole of the Earth's big internal magnet!

## Fig. 2 – World Magnetic Declination

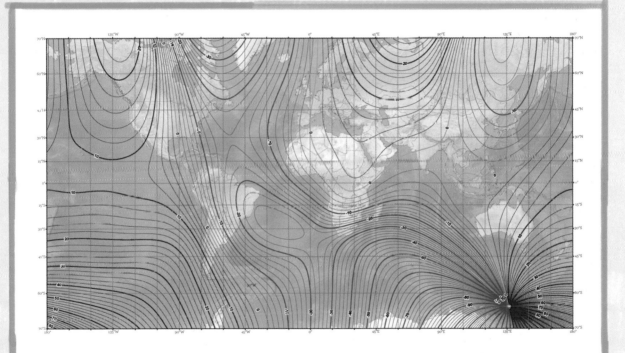

The isogonic lines on this map show the magnetic declination, or the amount magnetic north differs from true north, across the Earth. The lines are steadily changing because the magnetic poles move around 40km (25 miles) each year.

# PERPETUAL MOTION AND THE HEAT DEATH OF THE UNIVERSE

I have to admit to having a soft spot for perpetual motion machines, even though (or maybe because) they are so obviously doomed to failure. People have for centuries tried to build a machine that continues in motion without any input of energy, and perhaps able to export some instead. Their dreams of a better world, with free power and no pollution, have always, and must always, founder on the second law of thermodynamics.

The second law of thermodynamics tells us that the entropy of the Universe is always increasing. **Entropy**?

Entropy is sometimes considered a measure of disorder, easily illustrated by the bedroom of a typical teenager. It can be put into a higher state of order, if energy is put into the system, but rapidly reverts to a disordered mess.

A better way to think of entropy is as a measure of energy spread. A low-entropy situation is one where energy is concentrated – such as a cup of hot coffee. As it cools, it leaks energy to its surroundings and eventually will be the same temperature as the room, if you don't drink it first. The same thing happens when ice melts – the energy becomes more evenly spread, and entropy has increased.

Aha! I hear you say, but what about boiling a kettle, huh? That concentrates the energy and makes it less spread out! Doesn't this reduce the entropy of the Universe?

The energy to boil the kettle has not been magicked out of thin air, but has been generated in some power plant and transmitted to you. The actual energy source, whether it was hydroelectric, nuclear or fossil fuel, was relatively concentrated, and the act of harvesting and distributing it has caused it to become spread out. The local entropy in your kettle may have decreased, but the overall entropy of the Universe has still increased.

This, in essence, is why perpetual motion machines must always fail. Whatever initial energy is used to kick them into motion will inevitably become more spread out over time, through friction and air resistance and the generation of sound waves. All of those losses absorb mechanical energy and convert it to heat – the lowest grade of energy in the Universe.

Take this a bit further and we can see that, ultimately, every action eventually decays into lowly heat energy. Eventually, the whole Universe will have no starshine, no orbiting bodies, no sound or radio waves or anything, just heat. This is the end, the heat death of the Universe. Maximum entropy, when everything everywhere is at the same temperature and so no energy can flow.

The good news is that this is unimaginably far in the future, if it happens at all.

But perpetual motion machines? I'm afraid all they do is convert some useful high-grade energy into friction, slightly hastening the demise of all creation. So stop it, right now.

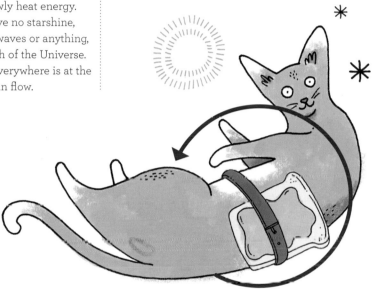

Toast lands buttered side down and cats always land on their feet: the buttered cat paradox is as close to perpetual motion as you're likely to get.

## The buttered cat

There is one perpetual motion mechanism that I'd really like to see in operation. It's well known that cats always land on their feet, and it is also a common observation that toast always lands buttered side down.

What if you strapped a slice of fresh buttered toast to a cat's back, and dropped it? The cat would twist to land cleanly, but the toast would exert a counter-force so that it reached the ground first. The result would be the two of them hovering just above the floor, spinning furiously in an attempt to win the battle.

The buttered cat paradox was created by John Frazee for *Omni* magazine, and makes me smile every time I think of it.

# THIS BOOK

There you are, minding your own business, reading a book about physics. Take a moment to consider some of the processes that have taken place, over fractions of a second and over billions of years, simply to place these few lovingly-crafted pages in your hand.

Deep in the heart of the Sun, atoms of hydrogen are stripped of their electrons and driven into each other at tremendous pressure and temperature. They fuse to form helium, releasing more energy as photons. It's a bit of a maelstrom in there, with these quanta of energy bashing through the fusion plasma soup, being re-absorbed and re-emitted. Ultimately photons are liberated from the Sun's surface, at the end of a process taking thousands of years, to begin the short flight (eight minutes twenty seconds, give or take) to Earth.

They do this with the sole purpose of plummeting through the atmosphere to land on this page and reflect back to your eye, so that you can read it. It's a popular activity among photons, with perhaps a quintillion ($10^{18}$) of them striking this page every second. Assuming, of course, that you are reading outside.

A few decades earlier, some identical photons escaped the Sun to be captured in the chlorophyll pigments of fast-growing softwood trees. Here they undertook their final quantum transformation, adopting a superposition to reach the cell's reaction centre via several routes at the same time. In the process, carbon dioxide from the atmosphere was combined with hydrogen to make carbohydrate that ultimately became the paper you are holding.

As you hold the book, the geometry of spacetime is distorted by the significant mass of the Earth in close proximity. This manifests as a gravitational attraction, accelerating the book toward the centre of the planet. You resist, mobilizing the awesome capabilities of your muscles to convert chemical energy from food you have eaten into mechanical energy in your arms and fingers.

More of this energy is expended when you turn the page, displacing air molecules and, if done vigorously enough, turning some of that energy into undirected sound. The tiny compression wave propagates through the nearby area, agitating some air molecules inside your ear which in turn stimulate nerve receptors to propel a small electrical signal to your brain.

In doing all this you respire, combining oxygen from the atmosphere with carbon from your food to produce carbon dioxide, replenishing some of what was absorbed by that tree earlier.

Every molecule, every atom, every proton, every quark, lepton and boson of what you are holding sprang into existence in the first second or so after the Big Bang. Their journeys to become part of this book involved many stages, with an overall theme of synthesis – the combination of smaller parts to make bigger and more complex ones.

Quarks make hadrons, which in turn make atoms. For most, their journey has gone no farther. Only about 4 percent of the Universe is regular matter, and of that 98 percent is just hydrogen and helium. Practically everything that you know, everything that you can see, feel and touch, is mostly made of some of the 0.08 percent of the Universe that is regular matter besides hydrogen and helium.

Elements in the upper part of the periodic table, including carbon and oxygen, are synthesized in relatively small stars like our own Sun, through the process of nuclear fusion. Heavier elements up to iron require the more intense pressure cooker of larger stars, while anything heavier still (copper, silver, iodine, for example) can only have been made in a stellar explosion – a nova or supernova.

And yet there are atoms and compounds of those things in and around you right now. Some will be in the paper, ink, binding and coating of this book.

What you are looking at is stardust, illuminated by star energy.

# INDEX

(page numbers in *italics* refer to figures and illustrations)

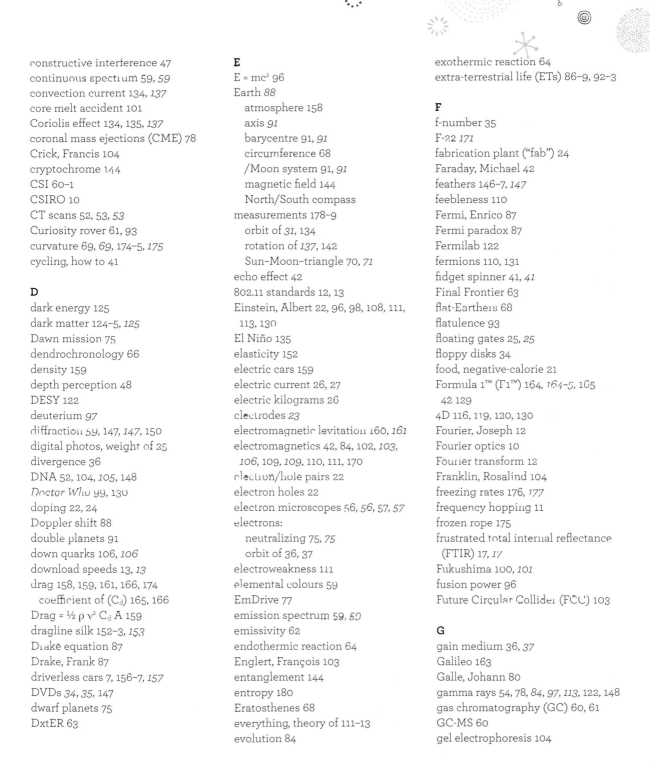

# PICTURE CREDITS

# ACKNOWLEDGEMENTS

This book would not have been possible without the steadfast support of my wife Mary, as well as the encouragement and occasional teasing of our children Freddie, Kate, and Theo. It is dedicated to them.

I would also like to thank my editor Polly Poulter for her patience and good humour when coping with my (mostly quite reasonable) demands, as well as illustrator and designer Tilly for bringing my words to visual fruition. It is humbling to be associated with such talent. Thanks also to publisher Trevor Davies, for his initial suggestion that I even think about doing this.

In the course of putting this book all together, I found myself consulting widely within the scientific community and seeking clarification from a number of experts, some quite eminent and others in their early careers. To all of them I offer this blanket acknowledgement, shamelessly intended to eliminate any risk of offending anyone by omission.

A number of personal and professional friends offered advice and the benefit of their own experience, notably Sarah Redsell, Phil Prime, David Bradley, Louis Barfe, and Guy Clapperton. It was appreciated.

# ABOUT THE AUTHOR

Russ Swan is an award-winning journalist who enjoys the challenge of complex technical topics. His work has appeared in *The Times*, *Financial Times*, *Guardian*, *Laboratory News*, *Wired* and *Gizmodo*, among other places. He has edited magazines on engineering, scientific research, and laboratory technology, and once caused the euro banknotes to be redesigned after pointing out that the pictures on them were copied from a book.

In his spare time he likes to take things apart and, sometimes, put them together again. This is why he currently has no working motorbike. He thinks his springer spaniel may have invented a portable wormhole generator, because of her ability to suddenly reappear at high speed from the opposite direction, and he loves bad science fiction.